中国敦煌历代服饰图典
The Illustration of China Dunhuang Costumes in History

『十四五』国家重点图书
『The 14th Five-Year Plan』of National Important Books

敦煌服饰文化图典
The Illustration of Dunhuang Costume Culture

盛唐卷（上册）
The High Tang Dynasty

Volume One

刘元风　赵声良　主编

Editors-in-Chief
Liu Yuanfeng
Zhao Shengliang

中国纺织出版社有限公司

·北京·

内 容 提 要

从书选择敦煌历代壁画（尊像画、故事画、经变画、史迹画、供养人像等）和彩塑中的典型人物形象，包括佛国世界中的佛陀、菩萨、弟子、天王、飞天、伎乐人，以及世俗世界中的国王、王后、贵族、平民等，对其反映的服饰造型和图案进行整理绘制，并对其文化内涵进行理论研究。

书中每一单元的内容包括敦煌典型洞窟的壁画或彩塑原版图片、根据此图像整理绘制的服饰效果图和重点图案细节图，以及重要图像的服饰复原图。书中还收录了与主题相关的学术论文，并对图像的历史背景、服饰特征、艺术风格等进行了深入研究和说明。

本书适合院校师生、科研人员、设计师和敦煌文化艺术爱好者学习借鉴，同时具有一定的典藏价值。

图书在版编目（CIP）数据

敦煌服饰文化图典. 盛唐卷. 上册 / 刘元风，赵声良主编. -- 北京：中国纺织出版社有限公司，2023.8
中国敦煌历代服饰图典 "十四五"国家重点图书
ISBN 978-7-5229-0622-5

Ⅰ.①敦… Ⅱ.①刘… ②赵… Ⅲ.①敦煌学—服饰文化—中国—唐代—图集 Ⅳ.① TS941.12-64

中国国家版本馆 CIP 数据核字（2023）第 094162 号

DUNHUANG FUSHI WENHUA TUDIAN
SHENGTANGJUAN SHANGCE

策划编辑：董清松　孙成成　　　责任编辑：孙成成
责任校对：王花妮　　　　　　　责任印制：王艳丽

中国纺织出版社有限公司出版发行
地址：北京市朝阳区百子湾东里A407号楼　邮政编码：100124
销售电话：010—67004422　传真：010—87155801
http://www.c-textilep.com
中国纺织出版社天猫旗舰店
官方微博 http://weibo.com/2119887771
北京雅昌艺术印刷有限公司印刷　各地新华书店经销
2023年8月第1版第1次印刷
开本：635×965　1/8　印张：39.5
字数：517千字　定价：598.00元　印数：1—1000册

中国纺织出版社有限公司
官方微博

中国纺织出版社有限公司
官方微信

编 委 会

Editorial Board

序

象征着中西方文化交流和友好往来的"丝绸之路",是古代人民用巨大的智慧创造出来的一条光彩夺目的美丽缎带,而敦煌莫高窟则是镶嵌在这条缎带上的一颗闪闪发光的明珠。在延续一千多年的敦煌石窟艺术中,各个历史时期的彩塑、壁画人物的服装和服饰图案非常丰富精彩,是我们取之不尽、用之不竭的艺术源泉。

我有幸自小跟随父亲常书鸿在敦煌学习壁画临摹,后来又因梁思成、林徽因两位先生走上了工艺美术的人生道路。父亲曾经提醒我:"不要忘记你是敦煌人……也是应该把敦煌的东西渗透一下的时候了!"多年来,我一直盼望对包括服饰图案在内的敦煌图案进行全面系统的研究。1959年,我和李绵璐、黄能馥两位老师曾去敦煌石窟专门收集整理服饰图案,这些原稿在20世纪80年代首先由香港万里书店出版为《敦煌历代服饰图案》一书。书中的图案部位示意图,是当时刚留校不久的刘元风老师和他的同学赵茂生老师帮忙绘制的。20世纪90年代,我又带领我的研究生团队对敦煌图案进行了分类整理,相继出版了《中国敦煌历代装饰图案》和《中国敦煌历代装饰图案(续编)》,完成了我的部分心愿。

2018年6月,敦煌服饰文化研究暨创新设计中心在北京服装学院正式挂牌成立。中心有两个主要任务:一是把敦煌服饰文化艺术作为专题进行深入化和系统性研究,二是根据研究成果和体会进行创新性设计运用,涵盖了继承与发展的永恒主题。围绕着这两个主要任务,中心举办并展开了学术论坛、科研项目、人才培养、设计展演等一系列具有学术高度且富有社会影响力的活动,获得了院校、科研院所及行业的广泛好评。

此次由北京服装学院和敦煌研究院两家单位合作推出的"中国敦煌历代服饰图典"系列丛书,集中了各自在敦煌学和服饰文化研究方面的专长和优势,在理论和实践两个方面进行继承和创新。一方面,以历史时代的划分进行敦煌石窟历史背景和服饰文化的理论阐述;另一方面,基于理论研究展开敦煌石窟壁画和彩塑服饰的艺术实践。丛书不仅以整理临摹的方式将壁画、彩塑的整体服饰效果和局部图案进行整理绘制,同时也在尊重服饰历史和工艺规律的基础上,将壁画和彩塑中的人物服饰形象进行艺术再现,使之更加直观生动和贴近时代。

对敦煌历代服饰文化的研究和继承工作,还仅仅是一个开始,绝不是结束。我希望更多的年轻学者、设计师、艺术家、学子们,能够继续前行,继续努力,做出更多的成绩,为弘扬我们引以为豪的敦煌艺术贡献自己的力量!

原中央工艺美术学院(现清华大学美术学院)院长

2020年9月

Preface

The Silk Road, which symbolizes the cultural exchange and friendship between China and the West, is a dazzling and beautiful ribbon created by the ancient people with their great wisdom. Mogao Grottoes is a shining pearl inlaid in this ribbon. It has lasted more than one thousand years, and the costumes and patterns on costumes of painted sculptures and mural figures in various historical periods are very rich and wonderful, which are our inexhaustible artistic source.

I was very lucky to learn mural copying at Dunhuang with my father Chang Shuhong when I was a child. Later, I took the road of arts and crafts because of Mr. Liang Sicheng and Mrs. Lin Huiyin. My father once reminded me: "Don't forget that you are from Dunhuang, you should study Dunhuang deeper..." For many years, I have been looking forward to a comprehensive and systematic study of Dunhuang patterns, including clothing patterns. In 1959, two scholars, Mr. Li Mianlu and Mr. Huang Nengfu and me went to Dunhuang Grottoes to collect and arrange clothing patterns. In the 1980s, these manuscripts were first published by Hong Kong Wanli Bookstore as the book *Costume Patterns from Dunhuang Frescoes*. The sketch illustration in the book was drawn by Mr. Liu Yuanfeng and his classmate Mr. Zhao Maosheng. In the 1990s, I led my postgraduate team to sort out Dunhuang patterns, and successively published the books *Decorative Designs from China Dunhuang Murals* and *Decorative Designs from China Dunhuang Murals(continued)*, which fulfilled part of my wish.

In June 2018, Dunhuang Costume Culture Research and Innovation Design Center was officially established in Beijing Institute of Fashion Technology. The center has two main tasks: one is to take Dunhuang costume culture and art as a special topic for in-depth and systematic research; the other is to carry out innovative design and application according to the research results and experience, covering the eternal theme of inheritance and development. Around these two main tasks, the center has held and launched a series of activities with academic height and social influence, such as academic forum, scientific research project, talent training, design exhibition and so on, which has been widely praised by colleges, research institutes and the industry.

The Illustration of China Dunhuang Costumes in History jointly launched by Beijing Institute of Fashion Technology and Dunhuang Research Academy focus on their respective expertise and advantages in Dunhuang study and clothing culture research, inherits and innovates in theory and practice. On the one hand, the historical background and costume culture of Dunhuang Grottoes are expounded theoretically according to the division of historical times. On the other hand, the art practice of Dunhuang Grottoes murals and painted sculptures costumes are carried out based on theoretical research. The series not only arrange and draw the overall clothing effect and patterns of murals and painted sculptures in the way of sorting and copying, but also on the basis of respecting the clothing history and technological rules. The series also make artistic reproduction of the characters' clothing images in the murals and painted sculptures, so as to make them more intuitive, dynamic and close to the times.

This research and inheritance work of Dunhuang costume culture is just a beginning, by no means the end. I hope that more young scholars, designers, artists and students could continue moving forward, working hard and making more achievements to contribute the promotion of Dunhuang art which we are proud of!

Chang Shana

Former president of Central Academy of Arts and Crafts

(now Academy of Fine Arts, Tsinghua University)

September 2020

目录

敦煌壁画与唐代时尚

赵声良

一、隋唐开放的国策与丝绸之路的繁荣

隋朝统一中国后，倾注了很大的精力来开通丝绸之路，目的就在于加强与西域的交往，以巩固政权和发展经济。大业五年（609年）隋炀帝西巡，经青海穿过祁连山而进入张掖，西域二十七国使节在张掖谒见炀帝，"皆令佩金玉，披锦罽，焚香奏乐，歌舞喧（諠）噪（譟）。复令武威、张掖士女盛饰纵观，骑乘填咽，周亘数十里，以示中国之盛"❶。这次盛大的外交活动，极大地促进了中国与西域的经济、文化交往，也为后来丝绸之路的持续繁荣打下了良好的基础。

隋末经过了一段时期的战乱，到唐初再度恢复丝绸之路的交通。唐在河西设凉、甘、肃、瓜、沙五州，又于天山南北置安西、北庭都护府，统辖轮台、伊吾、龟兹、于阗、疏勒、碎叶等镇，保证了丝绸之路的通畅。太宗时，又遣文成公主入藏，与吐蕃结盟，以保障青藏高原的稳定，使丝绸之路向南延伸。其后，王玄策经青藏高原，由尼泊尔进入印度，打通了中国与印度之间的吐蕃—尼婆罗道。唐代后期，由东南沿海地区到西方的海上交通发展起来，形成了海上丝绸之路，扬州等南方城市因此而繁荣。由于当时的航海技术还不完善，海上交通的危险性较大，因此，与国外的交流主要还是依靠陆路交通。唐代以后，形成了"伊吾之右，波斯以东，职贡不绝，商旅相继"❷的局面。中国的丝绸绢帛等纺织品源源不断地传输到西方，西方的骏马大量进入中国，满足了西方贵族的奢侈需求，也带来了东方马政的兴旺。良马的增加，加强了唐朝军队的实力，马在唐代政治中意义非凡，健壮的战马也成为那个时代男儿勇气与力量的象征，秦汉以来，丝绸之路的艺术往往与马有着密切的联系。武威出土的铜奔马，表现出剽悍的体态和优美的曲线，成为中国古代战马雕塑的代表。汉代以后，类似的奔马也出现在各地的壁画等艺术中。唐代的昭陵六骏，同样显示出唐代帝王对于勇健精神的向往。敦煌唐代壁画中也往往画出强健的马匹，反映了当时的审美倾向（图1）。

在佛教信仰繁荣的隋唐时期，伴随着佛教传播而带来的中外文化交流也十分发达。汉代以后，就有不少印度、西域的高僧，不远万里来到中国，传来了佛教文化（图2）。唐代的来华僧人见诸正史记载的就有十数人，在传播佛教的同时，也带来了印度和中亚的文化。而中国方面也不断有僧人西行求法。魏晋时代，就有法显、宋云西行求法。唐代玄奘西行取经（图3），辗转印度十多年，不仅从印度带来了大量

图1　莫高窟第431窟西壁　马夫与马

---

❶《隋书》卷67，《裴矩传》，北京：中华书局，1973年。
❷《新唐书·地理志》，北京：中华书局，1975年。

图2　敦煌绢画行脚僧图（唐，吉美博物馆藏）　　　图3　榆林窟第3窟　唐僧取经图（西夏）

佛经，而且把中国文化传播到了中亚和印度各地。在他西行与东返的过程中，还自然地承担了沟通往来和传播文化的使命，客观上加强了中国与西方各国各民族的联系。广大的中国人民因此而了解到印度和中亚各地的风情民俗，而沿途各地同样也因为玄奘而具体感受到了唐朝文化。除玄奘以外，唐代还有不少僧人或俗人往返于丝绸之路上，对中外文化的交流与传播作出过贡献。而在佛教传播的过程中，印度和西域的文化艺术也源源不断地传入了中国，中国的思想文化也同样反馈到印度和中亚、西亚各国。唐代，敦煌已成为丝绸之路上的一个重要佛教都会，当时莫高窟已有洞窟上千所。可以想见，东来西去的高僧大德在敦煌进行佛教活动是十分频繁的。因此，在当时宗教、文化交流的背景下，敦煌石窟的彩塑壁画所反映的应是当时流行的各种文化与艺术。

隋唐时期丝绸之路的繁荣，极大地促进了中外经济贸易的发展，中国的丝绸等纺织品、纸、陶瓷、金属工艺品等流传到了西方，而西域的金银器、玻璃、毛织物以及珍奇的动植物等也大量进入了中国。交流促进了繁荣，经济的发展带动了文化的兴盛，作为世界最繁荣的城市，唐代的长安也是引领时尚之都。在文学上，诗、赋流行，人才辈出。音乐、舞蹈、绘画、书法等艺术创意无限，乃至服饰、工艺等设计都不断求新求异，可以说唐代就是一个充满创新意识的时代。

二、盛唐之音

中国自古以来重视音乐文化，从先秦时代，就把礼与乐相结合，音乐就成为礼制的一个重要方面，形成了中国音乐文化的一个重要特征。因而，音乐实际上就成了贵族文化的象征，音乐演奏的排场，往往会体现着一定的等级。然而，随着丝绸之路的发展，中国与西方诸国的交流日益频繁，受外来文化的影响，音乐的平民化倾向日益明显，特别是在佛教法事活动中，往往会有大规模的音乐舞蹈表演。昔时贵族垄断的乐舞，早已进入百姓文化娱乐生活之中。而且，在外来音乐舞蹈的冲击下，传统的宫廷乐舞也产生了很大的变化，规模宏大的宫廷乐舞中，胡乐、胡舞竟占了相当大的成分。这是一个音乐文化因交流而繁荣的时代，也是一个音乐文化在民间普及的时代。

1. 隋唐宫廷音乐与寺院音乐

隋唐时期是中国音乐、舞蹈高度发达的时代，隋代宫廷设九部乐，唐初承隋制，后来增为十部乐，包括燕乐、清乐、西凉乐、天竺乐、高丽乐、龟兹乐、安国乐、疏勒乐、康国乐、高昌乐❶。这十部乐中，有四部为外国传入（天竺、高丽、安国、康国），有四部为国内少数民族音乐（西凉、龟兹、疏勒、高昌），只有燕乐与清乐为传统乐。这说明在丝绸之路文化交流中，大量的外国音乐传入中国，并深受中国人的喜爱，因而对这些音乐加以改编，并在国内流行开来。十部乐中分别配有不同人数的乐人，其中如规模较大的西凉乐配有乐器19种，乐工27人。此外，宫廷乐中又分出坐部伎和立部伎。特别是立部伎，规模宏大，演奏者多时达180人。除了宫廷音乐外，唐代民间的乐舞活动也十分普及，社会各阶层都有各种乐舞活动，民间的节日歌舞活动普及，在寺院还有与宗教活动相关的舞乐。

佛教寺院和石窟壁画中描绘有很多音乐、舞蹈形象。佛经中讲到对佛的供养有很多形式，其中音乐、舞蹈也是供养佛的方式，所以寺院和石窟艺术会出现很多音乐、舞蹈的形象。在很多佛教的活动中，诸如法会、行像以及佛教节日的活动，都会伴随着音乐、舞蹈的活动。佛教的发源地印度本来就是音乐、舞蹈十分发达的国家，在古印度早期的佛教雕刻中，如山奇大塔和巴尔胡特雕刻，就已出现了以音乐、舞蹈礼佛供养的场面，这样的习俗也随着佛教的传播而传入中国。在龟兹（今新疆库车）的佛教石窟壁画中就可看到大量天宫伎乐的形象，表现佛国世界的天人在演奏乐器或随着音乐的节奏而起舞。敦煌早期的壁画中，也有类似龟兹壁画的天宫伎乐形象。隋唐以后，表现佛国世界的经变画十分流行，经变中通常都要画出音乐、舞蹈的形象，以表现对佛的供养。同时，在经变画表现的佛国世界中，歌舞升平的美好场景也是不可或缺的。敦煌壁画中的音乐、舞蹈形象，与印度的壁画、雕刻中的乐舞有很大的差别：一是人物面貌，大多画成了中国人的形象；二是在服装上也有很大的不同，表现的应该是中国式的乐舞伎形象（图4）。联系起经变画中以中国式宫廷建筑来表现佛国世界，乐舞也同样是以中国乐舞的形象来表现佛国世界的景象，可以说，敦煌壁画中的音乐、舞蹈图像从一个侧面较真实地反映了隋唐时期我国音乐、舞蹈发展的盛况。

唐代经变画主要是表现净土世界的，如阿弥陀经变、观无量寿经变、药师经变等，在表现佛国世界场面中，都画出大规模的乐队和舞伎的形象，为我们展示了唐代音乐、舞蹈的盛况。乐舞场面的基本形

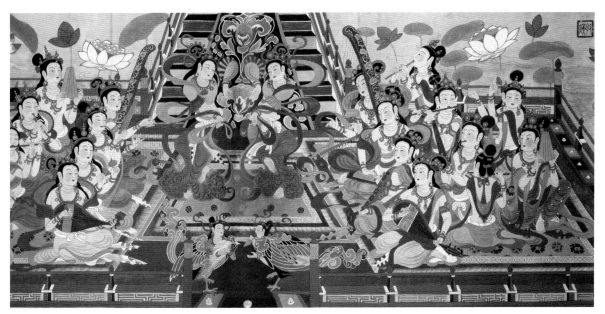

图4　莫高窟第148窟　药师经变中的乐舞图（万庚育复原临摹）

---

❶《旧唐书·音乐志》，北京：中华书局，1974年。

图5　莫高窟第220窟北壁　药师经变中的乐队局部（初唐）

式是中央有舞者起舞，两侧有乐队伴奏。乐队人数少则七八人，多则二三十人。例如，莫高窟第148窟东壁观无量寿经变中的乐队组成多达三十人，是乐队人数较多者。根据对壁画中乐队配器的调查，研究者认为敦煌壁画中的音乐总的来说是倾向于"西凉乐"的特征。《旧唐书·音乐志》记载西凉乐有"钟一架、磬一架、弹筝一、筝一、卧箜篌一、竖箜篌一、琵琶一、五弦琵琶一、笙一、箫一、筚篥一、小筚篥一、笛一、横笛一、腰鼓一、齐鼓一、鼓一、铜钹一、贝一、编钟今亡"❶。如果对照初唐第220窟药师经变画中乐器的配置，除了无钟、磬，却有方响外，其他各种乐器都有，而鼓的种类更多，总演奏人数达二十七人（图5）。虽然不完全符合史书所载的西凉乐，但大部分乐器都具备。在乐队中，往往打击乐的鼓排在较前列，方响、箜篌等大型的乐器会安排在显著的位置。

壁画中乐队的排列当然主要从画面的视觉上考虑，画家不一定按演奏时的真实情况表现。但即使是真实的演出，也同样要考虑观众的视觉感受，所以，壁画较真实地反映了当时的音乐演奏情景。

从壁画中还可看到隋唐时期各种乐器的形象，其内容涵盖了打击乐器、弹拨拉弦乐器、气鸣乐器等所有传统乐器的类型，可以说是一个古代乐器形象的博物馆。

2．巾舞、鼓舞、琵琶舞

隋唐以来，在开放与交流的形势下，外国和中国西部各少数民族的舞蹈在中原流行起来。从敦煌壁画中的舞蹈形象可看出巾舞、鼓舞、琵琶舞等舞蹈形象❷。"长袖善舞"是中国先秦以来的传统，汉代有巾舞，汉代绘画与雕塑中就有不少长袖舞形象。到隋唐时期，吸收了印度、中亚等外来形式，巾舞更加流行。例如，盛唐第172窟北壁观无量寿经变画中，二人持长巾起舞，一人表现正面，另一人为背面，两人均双手持飘带上举，长长的飘带绕过舞者的背部，从双臂而下，随着舞蹈动作，手持飘带上下回环，使飘带也形成美丽的弧线，增强了舞蹈的表现力。同样的双人舞在第148窟东壁观无量寿经变中也有出现（图6），也是一人为正面，另一人为背面，舞者所持飘带已形成一个个圆圈，显然舞蹈动作是十分急速的。第341窟的双人巾舞则表现出飘带随身体旋转的动态，说明舞者自己在持巾旋转。此外，第201窟和第112窟的巾舞则为单人舞，动作较舒缓悠扬。巾舞是敦煌壁画中出现较多的舞蹈形象，舞者持长长的飘带起舞，从飘带旋转起伏的形象，我们可以感受到舞蹈动作的前后轨迹。

还有不少舞者系腰鼓而起舞，这也是受西域影响而形成的舞蹈形式。榆林窟第25窟观无量寿经变中的舞蹈较为典型，舞者单腿而立，肩披长巾，神情专注，双手张开似乎正要奋力拍击腰鼓（图7）。类似的形象在莫高窟第108窟、156窟等窟中也可看到。腰鼓舞是一边击鼓一边舞蹈的，可以想见这是节奏鲜明、令人振奋的舞蹈艺术。除了腰鼓舞外，还有琵琶舞也是舞者边奏乐、边舞蹈的形象。琵琶舞既有

---

❶《旧唐书·音乐志》，北京：中华书局，1974年，第1068页。
❷ 王克芬、柴剑虹《箫管霓裳——敦煌乐舞》，兰州：甘肃教育出版社，2007年，第11–30页。

图6　莫高窟第148窟　舞伎（万庚育复原临摹）

怀抱琵琶而舞的，也有反弹琵琶而舞的。反弹琵琶可能是当时舞蹈的绝技，舞者左手高举将琵琶反置于背后，右手反向在后弹拨。在第172窟、112窟、156窟等洞窟中都有反弹琵琶舞的形象，说明这一造型是当时舞蹈中较流行的。第112窟的反弹琵琶舞尤为典型（图8），舞者上身半裸，项饰璎珞，下着锦裙，左腿站立，右腿高高提起；左手举起琵琶置于头后部，右手向后扬起作弹奏之姿；身体随着重心向右倾斜，两侧翻卷的飘带更加强了舞者的曲线造型。这个反弹琵琶的造型今天已成为敦煌壁画中舞蹈形象的代表。在很多双人舞场面中，或者鼓舞与琵琶舞，或者巾舞与琵琶舞相配合而舞，展示了唐代舞蹈精彩而丰富的场面。这些不同的舞蹈形象对于今天的舞蹈艺术创作来说，是取之不尽的灵感源泉。

敦煌壁画中多在经变画中涉及音乐、舞蹈，表现的是佛国世界的乐舞场景。其中不免有一些想象的成分，但还有一些是表现俗人的乐舞生活，这些场景更具有现实意义。盛唐第23窟北壁经变画中还绘出了一个拜塔舞乐的场面，在座方塔前，一组乐人在方毯上演奏音乐，一人面对佛塔跳舞，侧面又有一人虔诚地跪在塔前礼拜。这是表现当时以音乐、舞蹈礼佛的场面。在佛教信仰渗透到日常生活中的时期，音乐、舞蹈等活动也与佛教密不可分了。在经变画、故事画中表现世俗生活场景时，也常常画出一些单

图7　榆林窟第25窟南壁　腰鼓舞（中唐）

图8　莫高窟第112窟南壁　乐舞图（中唐）

独的乐舞场面。例如，盛唐第445窟弥勒经变中的嫁娶图，在宴饮场面中，有穿红衣的舞者挥袖起舞（图9），旁边有几个人弹奏乐器，真实地再现了普通民众生活中的乐舞场面。晚唐第85窟南壁报恩经变中表现的则是流落异国双目失明的善事太子在树下弹筝，而公主在旁静听的场面，十分生动。

图9　莫高窟第445窟　婚宴中的乐舞

三、服饰与流行色

唐代开放的社会环境，促成了服装时尚的发展，敦煌壁画中丰富的服饰形象，成为我们认识这个时代服装时最丰富的资料库。

1. 帝王及大臣的服饰

敦煌壁画在表现世俗人物听佛说法或听菩萨们辩论时，常常要画出帝王及庶民的形象，以表明世俗各阶层的人物对佛的虔诚。例如，莫高窟第220窟维摩诘经变中的帝王图（图10），皇帝戴冕旒，着冕服，青衣朱裳，曲领，白纱中单，并配大带大绶。衣上有日、月、山、川等所谓"十二章"的纹样，这是自先秦以来礼制所定的，只有天子的服装上才能出现表现天地山川的图案。大绶画出龙的形象，所有这些都符合当时的服制。从中，我们可以对史籍上说明的天子冕服有了明确的认知。周围大臣们的服装多着绛纱单衣，白纱中单，头戴进贤冠，足蹬笏头履，是唐朝官吏的朝服。在帝王两侧的大臣冠上还饰貂尾，唐代典籍所记"侍中、中书令、左右散骑常侍，则加貂蝉"❶，这样，我们就知道在帝王左右两侧的人当为侍中等职的人物。壁画中的帝王神态雍容，大臣们前呼后拥，可以与传为阎立本绘的《历代帝王图》（图11）相对照。

阎立本作为宫廷画师，有机会亲眼见到当时的皇帝、百官等情况，他的画当然能够比较真实地反映

图10　莫高窟第220窟东壁　帝王图（盛唐）

图11　传阎立本《历代帝王图》（波士顿美术馆藏）

---

❶《大唐开元礼》卷三，《序例下·衣服》，汲古书院，1972年，第30页。

皇帝、大臣等人物的服饰特征。所以，敦煌壁画唐代帝王图与阎立本作品的一致性，也表明了其真实可靠的程度。第220窟有贞观十六年（642年）题记，与阎立本生活的时代同期，说明阎立本在长安一带创制了"帝王图"之后，很快就在全国传播开了，乃至西北的敦煌也能得到摹本。类似的帝王及大臣的形象在初唐第335窟（686年）、盛唐第103窟、盛唐第194窟等窟都可以看到。在维摩诘经变中，大体形成了固定的模式，一直影响到晚唐、五代壁画。

2. 普通官吏及庶民服饰

魏晋以后，受胡服的影响，男子多穿窄袖长袍，但直到隋代，袍服往往较短，称为"大褶衣"，或许是受胡服袴褶的影响，具有紧身、窄袖、束腰的特点。从第62窟、281窟的隋代男供养人像上可以看出袍服稍短、露出腿部长靴的服饰特点。到了唐代以后，袍服就较长了，袍长至踝，在膝下部分加横栏，而称"襕衫"或"襕袍"。在盛唐第31窟、45窟、103窟、217窟等壁画中表现了普通男子形象（图12），就可看到作为男子日常服装的袍服，比隋代的服装加长了。庶民穿的袍往往较短，依然在膝盖以下露出靴子，并且在两侧开衩，以利于劳作。

官吏穿襕袍，往往腰系革带，着皮靴。这是唐代改革服制以后较为统一的样式。古代帝王和贵族所穿正规的鞋称"履"或"舄"，如"方头履""笏头履""云头履"等。汉晋以来墓葬多有出土，高级的一般有木底，鞋头部有装饰，较笨重。履的穿着与汉式的宽袍大袖服装相配合，表现贵族的气度，往往在礼仪场合使用，在实际生活中恐怕行动不太自由。而靴是以皮革制成，类似现代的高勒皮靴，最初是北方游牧民族所用，隋唐时期采用胡服，官员都用皮靴。在西安附近出土的壁画墓中同样也可以看出唐代男子穿袍服、系革带、戴幞头、穿皮靴的形象。

在很多洞窟经变画中的世俗人物或供养人像中，男子多为头戴幞头、身穿窄袖长身袍、足蹬乌皮靴的形象，这是官吏与士人的常服。盛唐以后，男子多穿襕袍，如第130窟的晋昌郡都督乐庭瑰供养像，就头戴幞头而身穿浅青襕袍（图13）。

幞头，是隋唐以来男子流行的首服，最初以幅巾包住头发，而在前后扎住，隋代壁画中供养人（第281窟）的幅巾前后各有两脚，与史书记载北周武帝"裁为四脚"[1]一致，还保持着幅巾的特点，这种形式的幞头到唐前期还存在，但幞头前部的两脚逐步变小，而后部的两脚加长。初唐以后，马周向唐太宗建议改革了幞头的形式，在全国推广，从敦煌壁画中可以看出唐前期的幞头在后面垂两脚已成为固定形式，是当时男子常服的一种。幞头的形式逐渐统一起来，头上分阶梯，前低后高，后部垂两脚。唐代出现了巾子（网帻），即以较硬之物做出形，固定住头发，然后在其上用幅巾扎成幞头的形

图12　莫高窟第103窟　拜塔图中的男子形象

图13　莫高窟第130窟　晋昌郡都督乐庭瑰供养像
（段文杰复原临摹）

---

[1]《资治通鉴》第172卷，"周主初服常冠，以皂纱全幅向后襆发，仍裁为四脚……"北京：中华书局，1956年，第5386页。

式。这样幞头就有了一定高度。武则天时期幞头高而前倾，称为"武家诸样"，玄宗时则多为圆顶。至唐后期，则完全做成类似帽子的形式，直接戴在头上，不再每天扎裹。幞头后部的垂脚长与短、软与硬，又有着时代的区分。唐前期大体是软脚，有的垂得较长，唐后期出现了宽而硬的垂脚，至五代发展为两侧伸出硬脚的幞头，宋代更加延长，成为展脚幞头。从敦煌壁画中可以看出幞头形式演变的完整轨迹。

3. 武士服饰

从彩塑和壁画中天王的形象，可以看出当时武士的甲胄装备，如初唐第322窟的天王像，头上的兜鍪，环项的顿项，上半身的掩膊、背韝、胸甲、身甲，下半身的战裙、行縢、毡靴等；盛唐第113窟的天王穿戴护项、披膊、兽头护肩、护臂、明光铠、束甲带、护腹、腿裙、皮靴等。这一系列装束都与《唐六典·武备志》记载相符合，说明这些天王服饰是按照当时的军人装备来塑造的。唐代多用铁甲，以长方形的铁片或铜片连缀而成，从壁画中表现的军士形象上就可以看到当时的铠甲形式（如明光铠）。第130窟东壁的练兵图，左侧一人骑马射箭，右侧军士数人在观看，他们均着方形铁片缀成的铁甲，与史书所载一致（图14）。此外，武将的常服还有战袍、衩衣等，头上还饰抹额。这些在壁画中都分别有所体现。

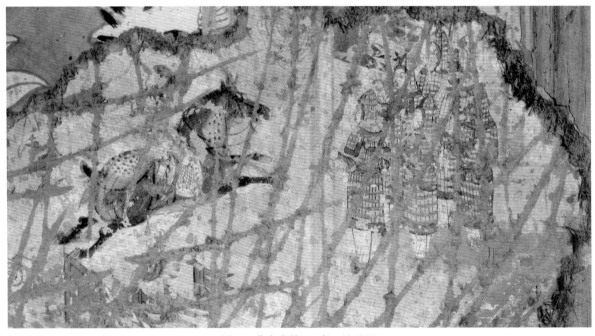

图14　莫高窟第130窟　练兵图

4. 妇女服饰

六朝至隋代，妇女多服襦裙，上半身紧窄，下半身裙裾曳地，反映出崇尚清瘦苗条的审美风尚。隋代妇女还流行身披披帛。披帛搭于肩上，或垂体侧，或绕臂而下，有一种飘逸的动感。唐代以后，随着不断扩大的对外开放，妇女服装时尚花样翻新，层出不穷。窄衫小袖和半臂，可以说是唐前期妇女的时尚，第329窟东壁的女供养人像，虔诚地跪在方毡上，双手持莲花花蕾作供养（图15）。她的上半身穿窄衫小袖，袖口很窄，圆领开得很低，裙子束腰很高。第217窟东壁画两妇女形象，其中后部一人可看出穿着如现代的短袖上衣，就是半臂（图16）。半臂本是庶民特别是劳动者所穿，唐代贵族妇女也往往以此为时尚，如长安附近礼县出土的新城长公主墓（663年）壁画中，就可看出公主及宫女们穿这样的半臂，但贵族妇女往往在窄袖衫上再穿半臂，这样就不会露出手臂的肌肤。

贵族妇女在正式场合多着袖口较宽的大袖襦，有披帛，下身则为曳地长裙。例如，第9窟的女供养人就是这样的服装，其服饰的质地十分讲究，花纹华丽，可见染缬的特点，从花纹图案上看，有单色、复色的"蜡缬"，还有"撮晕缬"，表现出唐代染缬技术的高度成就。

图15　莫高窟第329窟东壁　女供养人像

图16　莫高窟第217窟东壁　穿半臂的妇女
（段文杰复原临摹）

第130窟的都督夫人供养图中（图17），夫人头梳抛家髻，着碧罗花衫，外套绛地花半臂，穿红裙、云头履，披白罗花帔，一派雍容富贵的风度。跟随其后的女十三娘，头戴凤冠，斜插步摇，面饰花钿，着半臂衫裙、小头鞋履。

5. 妇女发式、面妆、头饰

唐代妇女的化妆有多种讲究，如以朱粉涂面被称作"红妆"。唐前期的红妆主要有两种，一种以朱红晕染额头及上眼睑，即"晓霞妆"，第332窟和第57窟的菩萨的形象就属于这一种；另一种是"黑眉白妆"，唐人记载当时的宫女们"施素粉于两颊，相号为泪妆"❶，在第329窟、220窟等壁画中的供养人和菩萨，以及第130窟的都督夫人一家的形象上均可以看到这样的妆容。唐代妇女画眉之风很盛行，初盛唐时期流行画长眉，称为娥眉，唐代诗人张祜写道："虢国夫人承主恩，平明骑马入宫门。却嫌脂粉污颜色，淡扫蛾眉朝至尊。"[《集灵台二首（之二）》]盛唐之后流行画短眉，如第130窟都督夫人供养图中的妇女全都画短眉（图18），与传为周昉的《簪花仕女图》中人物一样。

妇女脸上贴花的习俗很早，据说南朝宋武帝的寿阳公主在花园中假寐，梅花落于脸上，醒来后，脸上留下了梅花的印迹，宫女们觉得这样很美，竞相效仿，称为梅花妆❷。到了唐代，妇女在脸上贴花的种类就很多了。初唐壁画中已有妇女在额头上贴花钿的，正如唐诗所说"眉间翠钿深"❸。第130窟的都督夫人额上作五出梅花，脸上还有绿色花的面饰。晚唐五代以后，妇女脸贴花钿或画花子的风气一直很流行，在壁画中出现较多。榆林窟第25窟弥勒经变中，可见妇女在额头上画红花之形（图19），到晚唐、

---

❶《开元天宝遗事》卷下，北京：中华书局，2006年。

❷《太平御览》卷三十："宋武帝女寿阳公主人日卧于含章殿檐下，梅花落公主额上，成五出花，拂之不去。皇后留之，看得几时。经三日，洗之乃落。宫女奇其异，竞效之，今梅花妆是也。"（《太平御览》，中华书局影印本，第一册，1960年，第140页）

❸ 温庭筠《南歌子》，《温庭筠、韦庄词选》，上海：上海古籍出版社，2002年。

图17　莫高窟第130窟　都督夫人供养图（段文杰复原临摹）

图18　莫高窟第130窟　妇女面妆

图19　榆林窟第25窟　妇女头饰与面妆

五代时期，此风更浓。由于五代壁画中供养人像形体较大，表现得尤为清晰。例如，第98窟、61窟女供养人像，脸的上半部以胭脂画出半月形红晕，在其上又绘以花形。在脸颊之处又常常绘以凤鸟，或蝴蝶，或花朵之形，丰富无比。

唐代妇女的发式也十分讲究，唐前期多高髻，高髻又分高耸如椎的椎髻与侧向一面的半翻髻等。开元天宝之后流行抛家髻，脸两侧的头发垂下形成两鬓包面的形式，上部的头发则向上做成一定的造型，如椎形、花形等，莫高窟第130窟都督夫人礼佛图中的妇女大多为抛家髻。不仅敦煌壁画中多见，传为周昉的《挥扇仕女图》等唐代传世本绘画中，也可看到类似的发型。

唐前期妇女往往于头发上插花朵，或不加装饰。唐后期妇女头上的装饰渐多，特别是晚唐、五代时期，妇女头上插簪、插梳子等装饰物越来越流行，如晚唐第9窟供养人在额头上画花子，头上插满簪、花，多至十数件（图20）。直到五代以后，妇女头部的装饰更加复杂，面部贴花的形式也丰富多样。例如第98窟、61窟的女供养人就是代表，而其中又有回鹘公主等妇女头饰除了插簪外，还分别戴回鹘族的桃形冠和凤冠等，又有步摇为饰，无比华丽。

6. 唐朝的流行色

唐代文化是充满时尚感的，唐朝的流行色也是随着社会发展而不断翻新的。初唐石窟壁画，色彩清新，如春风温润，万物萌生。淡淡的青绿色调成为壁画中最流行的色调。建于贞观十六年的莫高窟第

图20　莫高窟第9窟　女供养人（欧阳琳、史苇湘复原临摹）

220窟保持了初创时的壁画色调，此窟壁画以明亮的青绿色调为主，西方净土变中以浅绿色表现水池，莲花则多以蓝色描绘，又有众多绿色的树木，形成十分协调的青绿色调，表现出明净优雅的气氛。同窟北壁的药师经变，表现的是东方净土的净琉璃世界，地面多用赭红色与白色花点交织，以表现琉璃、玛瑙等宝石，整壁的背景色依然是以青绿色为主调。七身立佛分别穿深赭色或土红色的袈裟，与背景的绿色以及地面的赭红色调相协调。可以说，第220窟壁画的色调是青绿色与赭红色交织的轻快而典雅的风格。同为初唐时期的莫高窟第321窟，南壁十轮经变以宏伟的青绿山水为背景，其中的人物服饰或为与背景一致的青绿色，或为艳丽的土红色与朱砂色，显示出鲜艳而明快的风格。此窟北壁表现西方净土变，主体画面是绿水池中楼阁与平台、栏杆、小桥等建筑，画面上部相当大的面积表现天空的景象，以深蓝色表现广阔的天空，空中有宝楼阁、宝树、不鼓自鸣的乐器以及来来往往的飞天等形象，显示出空中的明净。本窟西壁的佛龛顶的壁画也配合南北两壁的色调，表现深蓝的天空中自由飞行的飞天。此窟的色调以天空的深蓝色为主调，配合石绿色的山水等景物，表现出鲜明而高雅的风格。初唐洞窟还有一个类型是以土红色调为主的壁画，如莫高窟第57窟、322窟、328窟、329窟、335窟等。第329窟正面龛内壁画以红色调为主，表现佛传故事；窟顶四披均在土红底色中表现千佛；南北两壁整铺经变画均以土红赭色为主调，表现佛国世界的殿堂建筑，并辅以深蓝色与石绿色调表现天空和水池，整窟呈现热烈、昂扬的气氛。

进入盛唐，洞窟壁画的色调更加丰富，以第23窟、217窟、148窟等为代表的盛唐窟，壁画中青绿色调仍然是最为流行的基调。但此时的青绿色逐渐以石绿色为主调，比起初唐的青绿色显得更为沉着而厚重了。如果说初唐的青绿色体现出的是春意盎然，此时的青绿色则是盛夏的密林了。在青绿色流行的同时，赭红色调也有一点变化，就是红的地方更加鲜明而突出，与同样深重的蓝色、绿色以及部分黑色调并列在画面中，使画面显得更加富丽而深厚。第320窟、171窟、172窟等洞窟的壁画色彩在厚重、沉稳中也体现着丰富多彩、华美灿烂的倾向。

盛唐后期出现了一些色彩简淡的壁画，到了中唐以后就形成了流行倾向。这些洞窟的壁画色彩数量减少，设色淡雅，石绿色统摄着全部壁画，形成了唐代后期洞窟的另一种面貌，以莫高窟第159窟、196窟等为代表，石青色、石绿色都用得较淡，还多用白色或浅黄色，使洞窟总的风格偏浅淡清雅。

唐朝壁画中的世俗人物服装，也为我们提供了当时的人物服饰中的流行色。这些人物服饰所体现的倾向也大致与壁画的总色调一致。男子服饰多为圆领袍，变化不大，多为红色、赭色、黑色，也有白色与蓝色。总体来说，以深赭红色调为主。妇女服饰变化较多，初唐时期，可见粉红、浅黄、淡蓝色调的服饰，盛唐以后，灿烂多彩，衣纹的花饰繁多且层次变化丰富，显示出唐代纺织、染缬等方面技术的高度发达。同时，在唐朝壁画中的人物装饰中，宝石、玉器、琉璃（玻璃）、玛瑙等物也竞相绽放光彩，为这一时代的流行色增添了无限丰富的意趣。

总之，唐朝的流行色极为丰富，但有一些基本色调是贯穿整个唐代的，如青绿色系、赭红色系、黑褐色系。其中，初唐的色彩明亮绚丽，盛唐的色彩沉稳厚重，中晚唐的色彩婉约清丽。而在服装、建筑装饰、器物等方面的色彩应用也不尽相同，但在流行色的大趋势下，各方面的色彩也会相应地有所变化。

四、"胡风"与国际交流

唐代是一个十分开放的时代，以开阔的胸襟，广泛地接受外来文化，从而丰富了自己，强大了中华。隋唐时期西域各方文化流入，在中国就出现了胡服、胡帐、胡床、胡座、胡饭、胡箜篌、胡笛、胡舞等。正如鲁迅所说："唐室大有胡气。"[1]这一"胡化"现象，同样在敦煌艺术中体现出来。位于丝绸之

---

❶ 鲁迅《致曹聚仁（1933.6.18）》，《鲁迅书信集》上册，北京：人民文学出版社，1976年，第379页。

路要冲的敦煌，比起内地城市来，中外使节与客商的来往更加频繁，而且，还常常有多国、多民族的人聚居此地，中国古代与西域的交往，在这里留下了丰富的形象资料。

中国的历史记载，葡萄是张骞出使西域后从大宛带来的，《史记》还记载了大宛一带用葡萄做酒，人们都喜欢喝葡萄酒。据相关的历史研究，葡萄原产于小亚细亚里海与黑海之间及其南岸地区，波斯（今伊朗一带）和埃及是最早栽培葡萄和用葡萄酿酒的两个古国。汉代以后，葡萄与葡萄酒都传入了中国。从汉代到唐代的历史记载可知，葡萄的产地逐渐由西方的波斯向东发展，到唐代为止，西域的龟兹（今库车）、且末、于阗（今和田）、高昌（今吐鲁番）、伊州（今哈密）等地都种植了葡萄，并可酿造葡萄酒。《唐会要》等文献还记载唐太宗平高昌，引进了葡萄种植和葡萄酒酿造技术，此后，汉民族地区也能酿造葡萄酒了[1]。唐朝的葡萄酒产地主要有西州（今吐鲁番地区）、凉州（今甘肃省武威市）、并州（今山西省太原市一带）。尽管如此，葡萄和葡萄酒在唐代仍然是珍贵之物，天宝末的进士鲍防在诗中写有"天马常衔苜蓿花，胡人岁献葡萄酒"[2]，说明当时葡萄酒作为西域的方物来进贡，是珍贵之物。在唐代壁画和工艺品中，我们就可以看到很多葡萄的装饰图案（图21），说明当时人们对葡萄的喜爱。

狮子原产非洲和西亚，有的学者从"狮子"一名的读音，推测其来自波斯[3]。中国史书记载早在西汉时期，狮子已由西方国家输入中国。由于狮子凶猛，东汉以后常借用狮子的形象作为镇墓兽，称为"辟邪"。唐代以后，西域来献狮子的记载就更多了。例如《唐会要》记载贞观九年（635年）七月，康国"献狮子，太宗嘉其远来，使秘书监虞世南为之赋。"[4]此外，唐代还有西域的米国、吐火罗国等国献狮子的记录。狮子在佛教中具有神圣的地位，如文殊菩萨的坐骑就是狮子。所以，中国人往往把狮子看作神物。但当时的艺术家是可以见到实际的狮子，绘画中的狮子也应该是有依据的。壁画中不仅画出了文殊菩萨骑着狮子，还画出了牵狮的"昆仑奴"（图22）。"昆仑奴"肤色为棕色，卷发，可能来自古代东南亚一带。壁画中，普贤菩萨的坐骑白象也是由"昆仑奴"牵引的，这些来自外国的动物，由外国人物来牵也合乎常理。

大象在唐代也是外来的动物。虽然在青铜时代中国北方曾经有象生存，所以在商周的青铜器中我们可以见到不少有象的装饰形象，但随着时代气候的变迁，大象在北方终于绝迹。自佛教从印度传来之后，在佛教故事中有不少与象有关的内容，因为印度也是

图21　莫高窟第209窟　葡萄石榴纹藻井
（段文杰复原临摹）

图22　榆林窟第25窟　狮子和"昆仑奴"

❶ 海滨《唐诗所见葡萄和葡萄酒文化景观》，《西域文史》第三辑，2008年12月。
❷ 鲍防《杂感》，《全唐诗》卷307，北京：中华书局，1960年。
❸ 谢弗著、吴玉贵译《唐代的外来文明》，北京：中国社会科学出版社，1995年。
❹《唐会要》卷99，上海：上海古籍出版社，1991年，第2105页。

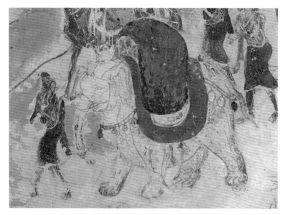

图23　莫高窟第103窟　胡人牵象

图24　莫高窟第217窟南壁　戴风帽的人物

图25　莫高窟第217窟东壁　胡服人物
（李其琼复原临摹）

大象较多的国家，于是大象的形象在佛教艺术中时有出现。唐代是中外交流空前发达的时代，外国的大象也来到中国，据唐代的史书记载，南方的林邑、真腊等国都曾向唐朝进贡大象。当然，这些大象到了中国，便成了珍稀动物，通常只是关在禁苑，遇到有庆典活动时，拉出来让士庶百姓观赏一下，增加热闹的气氛而已。尽管如此，与佛教密切相关的大象，总还是让中国人有了真实的感受。因此，在佛教艺术中画出的大象也就有了真实性（图23）。

隋唐时期，由于丝绸之路的通畅，与外国的交往很多，受外国的影响，所谓"胡服"也在汉地广泛地流行起来。例如，贵族男女出行时喜欢戴的"胡帽"就是效仿突厥和东伊朗人的服饰。唐前期妇女们常用的"幂䍦"就是把胡帽与面纱结合，遮蔽头顶、面部。"幂䍦"往往配合类似披风的外套使用，这样的服饰既有助于贵族妇女外出时遮风，又可免受外人窥视。后来改短，称为"帷帽"，在帽子下垂下帷，用以遮挡风沙，可以说是胡服的改革形式。在莫高窟第217窟壁画中就画出一个骑马穿红色披风的人物，头戴帷帽，在崇山峻岭中行进（图24）。

翻领大衣是西域胡服中流行的形式，在中国西部乃至中亚一带都较常见。现存的中亚粟特壁画和龟兹壁画中，就可看到着翻领大衣的粟特或龟兹人，这种大衣的面料较硬，可能为革制或者是较厚的毛织物。这样的服饰，在唐前期却成为中原汉族的时尚，在长安一带的壁画墓中可以看到贵族穿这样的服装，如陕西出土的韦浩墓（708年）壁画中就有不少穿胡服的人物，有的还是妇女穿胡服男装，这在当时也是较时尚的穿法。在敦煌壁画中也可看到男子穿胡服的形象，如第217窟东壁人物（图25），第445窟嫁娶图中的主要人物等。

敦煌与西域的少数民族的交往频繁，唐代聚居于敦煌的波斯人、龟兹人、回鹘人、吐蕃人都很多，汉族穿胡服、戴胡帽的形象也就不足为奇。而壁画中大量的外国人物或西域民族人物也为我们认识当时胡汉交融的历史提供了丰富的图像资料（图26）。

南北朝到隋唐时期，波斯的工艺品通过丝绸之路大量传入了中国，如狩猎图是古代波斯最流行的主题，表现狩猎形象的银盘就曾在中国北方发现很多。波斯萨珊王朝（3～5世纪）正是古罗马帝国强盛的时

图26　莫高窟第103窟　外国人物图

期，欧洲的古罗马文化、西亚的波斯文化与中国文化在从中亚到西亚辽阔的地域相碰撞，处处留下了文明交汇的脚印。

在隋到初唐时期敦煌彩塑与壁画的菩萨服饰中，就有大量的波斯纹饰，如联珠对鸟纹、联珠对兽纹、联珠飞马纹等。直到初唐时期，波斯风格的联珠纹依然盛行。在新疆一带出土的纺织物中，也有联珠纹图案的，最初这样的联珠纹图案是来自波斯的织物。但在唐以后，中国也能按波斯的技术来制作织锦了。所以在丝绸之路沿线出土的唐代织锦中，常常会发现其图案风格和制作方法完全是波斯式的，但其中又织出汉字纹样。联珠纹的流入对唐代装饰纹样产生了深远的影响，对于团形纹样的喜爱，逐渐形成了中国传统纹样的一个特点。

此外，源自中亚的葡萄纹、石榴纹也在敦煌石窟中广泛采用，并不断进行改造，与卷草纹相结合，形成丰富的装饰纹样，在初唐、盛唐的壁画中大放异彩（图27、图28）。

玻璃、琉璃，曾经是十分珍贵的东西，最初传入中国时只有帝王和一些贵族才能够拥有。所以在隋唐敦煌壁画中，常常画出菩萨手持玻璃器皿的形象，以显示其尊贵。玻璃器皿有碗形镶嵌宝珠的，有杯子形的，表现出当时的制作工艺。

香炉也是礼佛时少不了的供器，往往借鉴了外国的样式而制作出来。唐代的香炉工艺精湛，样式繁多。法门寺地宫出土的镀金银香炉，底座有六腿，

图27　莫高窟第23窟
石榴卷草纹

图28　莫高窟第444窟　葡萄石榴卷草纹头光

香炉盖为莲花花蕾形，底座和炉壁的边缘分别有坠饰。类似的香炉在敦煌壁画中也时常可见。唐代后期壁画中流行的香炉特点是，下部有莲花形底座，上部敞口呈八边形或六边形，上部有圆形镂空的盖，中央往往饰宝珠。

五、小结

综上所述，唐朝政治开明，经济高度发达，特别是丝绸之路的繁荣，极大地促进了中外文化的交流。外来文化与中国传统文化交流互鉴，为唐代的文化艺术鼎盛提供了条件。从时尚这个方面来看，唐朝的艺术充满了创造力，仅在敦煌壁画中所见，已是层出不穷，不断创新，令人目不暇接。限于篇幅，本文仅就以上几个方面作一探讨，管中窥豹而已。

Dunhuang Murals and the Fashion of the Tang Dynasty

Zhao Shengliang

1. The opening imperial policy in the Sui and Tang Dynasties and the prosperity of the Silk Road

After the unification of China in the Sui Dynasty, great efforts were devoted to opening the Silk Road, with the aim of strengthening contacts with the Western Regions in order to consolidate political power and develop economy. In the fifth year of Daye (609), Emperor Yang of Sui Dynasty visited the west part of China. He passed through Qilian Mountain and entered Zhangye via Qinghai province. Envoys of 27 countries in the Western Regions were invited to meet him in Zhangye. "They were all required to wear gold, jade and brocade, burn incense and play music, and sing and dance loudly. The emperor also ordered men and women in Wuwei and Zhangye dress up and go outside, rides and carriages caused traffic congestion for tens of miles to show the prosperity of China." This grand diplomatic event greatly promoted the economic and cultural exchanges between China and Western countries, and also laid a good foundation for the continued prosperity of the Silk Road.

After a period of war in the late Sui Dynasty, the traffic on the Silk Road was restored again in the early Tang Dynasty. In the Tang Dynasty, five prefectures of Liang, Gan, Su, Gua and Sha were set up in Hexi region, and Anxi and Beiting Protectorate were set up in the north and south of Tianshan Mountain, governing Bugur, Yiwu, Kucha, Khotan, Shule and Tokmok, ensuring the smooth passage of the Silk Road. During the reign of Emperor Taizong, Princess Wencheng was sent to Tibet to form an alliance with Tubo to ensure the stability of Qinghai-Tibet Plateau and extend the Silk Road to the south. Later, Wang Xuance entered India from Nepal via Qinghai-Tibet Plateau, opened up the Tubo-Nepal road between China and India. In the late Tang Dynasty, the maritime trade from the southeast coastal areas to the west developed, forming the Maritime Silk Road, so that Yangzhou and other southern cities prospered. Because the navigation technology was not good enough at that time, the risk of maritime traffic was great. Therefore, the communication with foreign countries mainly depended on land traffic. After the Tang Dynasty, the situation of "the right of Yiwu, the east of Persia, continuous tributes, and successive business trips" was formed. Chinese textiles such as silk were continuously transported to the west, and a large number of western horses entered China, which not only satisfied the luxury of western nobles, but also brought the prosperity of oriental horse breeding. The increase of good horses strengthened the strength of the army of the Tang Dynasty, and horses had great significance in the politics of the Tang Dynasty. Strong war horses also became a symbol of man's courage and strength at that time. Since the Qin and Han Dynasties, the art on the Silk Road has often been closely related to horses. The Galloping Bronze Horse unearthed in Wuwei shows vigorous posture and beautiful curves, and has become a representative of ancient

Fig. 1 Groom and horses on the west wall in Cave 431 at Mogao Grottoes

Chinese war horse sculpture. After the Han Dynasty, similar galloping horses also appeared in murals and other arts everywhere. The Six Steeds of Zhao Mausoleum of the Tang Dynasty also showed the emperor's longing for the spirit of courage and health. Strong horses are often painted in Dunhuang murals during the Tang Dynasty, reflecting the aesthetic tendency at that time (Fig. 1).

In the Sui and Tang Dynasties, when Buddhism flourished, the cultural exchanges between China and foreign countries brought by Buddhism were also very developed. After the Han Dynasty, many eminent monks from India and Western Regions traveled thousands of miles to China and spread Buddhist culture (Fig. 2). There were more than ten monks who came to China in the Tang Dynasty. While spreading Buddhism, they also brought India and Central Asia culture. On the Chinese side, monks had continued to travel westward to seek Dharma. In the Wei and Jin Dynasties, Fa Xian and Song Yun went west to seek Dharma. Xuanzang in the Tang Dynasty traveled westward to learn scriptures (Fig. 3) and traveled in India for more than ten years. He not only brought a large number of Buddhist scriptures from India, but also spread Chinese culture to Central Asia and India. During his journey to the west and return to the east, he naturally undertook the mission of communication and cultural dissemination, and objectively strengthened the ties between China and various nationalities in the Western Regions. As a result, the vast number of Chinese people learned about the customs and folk cultures about India and Central Asia, and all places along the way also experienced the culture of the

Fig. 2 Dunhuang silk painting of walking monks, Tang Dynasty, Musée Guimet Collection

Fig. 3 Xuanzang journey to the west, Cave 3 of Yulin Grottoes, Western Xia

Tang Dynasty because of Xuanzang. In addition to Xuanzang, many monks or laymen in the Tang Dynasty traveled back and forth on the Silk Road and made contributions to the exchange and dissemination of Chinese and foreign cultures. In the process of Buddhism spread, the culture and art of India and the Western Regions had also been continuously introduced into China, and China's ideology and culture have also been fed back to India, central and west Asian countries. Dunhuang had become an important Buddhist city on the Silk Road in the Tang Dynasty. At that time, there were already thousands of caves at Mogao Grottoes. It is conceivable that eminent monks from east to west carry out Buddhist activities in Dunhuang very frequently. Therefore, under the background of religious and cultural exchanges at that time, the painted murals in Dunhuang Grottoes should reflect various popular cultures and arts at that time.

The prosperity of the Silk Road in the Sui and Tang Dynasties greatly promoted the development of Chinese and foreign economy and trade. China's silk and other textiles, paper, ceramics and metal handicrafts spread to the west, and a large number of gold and silver ware, glass, wool fabrics, rare animals and plants from the Western Regions also entered China. Communication promotes prosperity, and economic development drives the prosperity of culture. As the most prosperous city in the world, Chang'an in the Tang Dynasty was also a capital of fashion. Literature, poetry and Fu were popular, and talents emerged in large numbers. Music, dance, painting, calligraphy and other artistic activities had unlimited creativity, and even clothes, crafts and other designs were constantly seeking innovation and difference. It can be said that the Tang Dynasty was an era full of innovative habit.

2. Music of the high Tang Dynasty

Since ancient times, China has attached great importance to music culture. Since the pre-Qin period, the combination of ritual and music has become an important aspect of ritual system and formed an important feature of Chinese music culture. Therefore, music has actually become a symbol of aristocratic culture, music performance pomp often reflects social status. However, with the development of the Silk Road and the contact between China and the Western countries, influenced by the foreign culture, the trend of popularizing music was becoming more and more obvious, especially in Buddhist ritual activities, there were often large-scale music and dance performances. In the past, music and dance monopolized by aristocrats, and now it has already entered people's cultural and entertainment life. Moreover, under the impact of foreign music and dance, the traditional court music and dance had also undergone great changes. In the large-scale court music and dance, foreign music and foreign dance account for a considerable proportion. This was not only an era when music culture was prosperous due to communication, but also an era when music culture was widely popularized among people.

2.1　Court music and temple music in the Sui and Tang Dynasties

During the Sui and Tang Dynasties, Chinese music and dance were highly developed. In the Sui Dynasty, there were nine music styles in the court, following the Sui system to the early Tang Dynasty, then increased to ten music styles, including Yan music, Qing music, Xiliang music, Tianzhu music, Gaoli music, Qiuci music, Anguo music, Shule music, Kangguo music and Gaochang music. Among them, four were imported from foreign countries (Tianzhu, Gaoli, Anguo and Kangguo), and four were brought from domestic minorities

(Xiliang, Qiuci, Shule and Gaochang). Only Yan music and Qing music are traditional music. This shows that in the cultural exchange along the Silk Road, a large number of foreign music was introduced into China and was deeply loved by Chinese people. Therefore, these music were adapted and became popular in China. There are different numbers of musicians in each of the ten styles. For example, the large-scale Xiliang music has 19 kinds of instruments and 27 musicians. In addition, court music can be divided into sitting musicians and standing musicians. In particular, standing musicians has a large number even reach to 180 performers sometime. In addition to palace music, folk music and dance activities in the Tang Dynasty were also very popular. There were various music and dance activities at all levels of society, folk festival singing and dancing activities, and dance and music related to religious activities in temples.

There are many music and dance images depicted in the murals of Buddhist temples and grottoes. The Buddhist scriptures say that there are many ways of offering for Buddha, among which music and dance are also the way to offer for Buddha, so there will be many images of music and dance in temple and grotto art. In many Buddhist activities, such as Dharma assembly, statue worship and Buddhist festivals, music and dance activities will be accompanied. India, the birthplace of Buddhism, is a country with very developed music and dance system. In the early Buddhist carvings in ancient India, such as Sanchi Monument and Barhut carvings, there have been scenes of worshiping Buddha with music and dance. Such customs have also been introduced into China with the spread of Buddhism. A large number of images of heaven musicians and dancers can be seen in Buddhist grotto murals in Qiuci (now Kuqa, Xinjiang), showing that people in Buddhist world are playing musical instruments or dancing with the rhythm of music. In the early Dunhuang murals, there are also heaven musicians and dancers images similar to the Qiuci murals. After the Sui and Tang Dynasties, sutra illustration painting to show the Buddhist world was very popular. In sutra illustration, the image of music and dance was usually painted to show offering to Buddha. At the same time, in Buddhist world represented by sutra illustration, the beautiful scene of singing and dancing is also indispensable. The music and dance images in Dunhuang murals are very different from those in Indian murals and sculptures: first, the characters are generally painted as the images of Chinese people; second, there are great differences in clothing, which show the image of Chinese musicians and dancers (Fig. 4). Connecting with the Buddhist world illustrated by Chinese palace architecture, music and dance also in the image of Chinese music and dance. It can be said that the music and dance images in Dunhuang murals truly reflect the grand development of music and dance in China during the Sui and Tang Dynasties.

The sutra illustration paintings in the Tang Dynasty mainly show the Pure Land World, such as Amitabha Sutra illustration, Amitayurdhyana Sutra illustration, Bhaisajyaguru Sutra illustration, etc. In the illustration of Buddhist world, they all have the images of large-scale bands and dancers, showing us the grand scene of music and dance in the Tang Dynasty. The basic form of music and dance scene is that there are dancers in the center and bands on both sides. The members of the band are as few as seven or eight, as many as twenty or thirty. For example, the band in the Amitayurdhyana Sutra illustration on the east wall of Cave 148 at Mogao Grottoes consists of up to 30 people, which is one with a large number of members. According to the investigation of the band instruments in the murals, researchers believe that the music in Dunhuang murals generally close to the characteristics of "Xiliang music". According to *Old Tang Book-Music*, Xiliang music includes "a bell, a qing, a pluck zither, a zither, a lying harp, a vertical harp, a pipa, a five-string pipa, a sheng, a xiao, a bili pipe, a small bili pipe, a flute, a horizontal flute, a waist drum, a qi drum, a drum, a copper cymbal,

Fig. 4 Music and dance scene in the Bhaisajyaguru Sutra illustration of Cave 148 at Mogao Grottoes (Copied by Wan Gengyu)

a shell, no chime now days". Compared with the configuration of musical instruments in Bhaisajyaguru Sutra illustration in Cave 220 of the early Tang Dynasty, there are all kinds of musical instruments that mentioned in the records except bell and qing, but has fangxiang, and there are more types of drums, with a total number of 27 players (Fig. 5). Although it is not completely consistent with the Xiliang music recorded in historical books, most musical instruments have included. In the band, the drums of percussion music are often arranged in the front row, and large musical instruments such as fangxiang and konghou will be arranged in prominent positions. Of course, the arrangement of bands in the mural was mainly considered from the visual aspect of the picture, and the painter didn't have to arrange according to the reality. However, even if it is a real performance, the visual feeling of the audience should also be considered. Therefore, the murals possibly still truly reflect the music performance information at that time.

From the mural, we can also see the images of various musical instruments in the Sui and Tang Dynasties. Its content covers all types of traditional musical instruments such as percussion instruments, pluck string instruments and blowing instruments. It can be said that this is an ancient musical instruments' image museum.

Fig. 5 The music band in Bhaisajyaguru Sutra illustration (part) on the north wall of Cave 220 at Mogao Grottoes dated to the early Tang Dynasty

2.2   Ribbon dance, drum dance, pipa dance

Since the Sui and Tang Dynasties, under the environment of openness and exchange, the minority dances in western China and foreign dances have become popular in the Central Plains. From the dance images in Dunhuang murals, we can see dance images such as ribbon dance, drum dance and pipa dance. "Long sleeves and dance well" has been a tradition in China since the pre-Qin Dynasty. In the Han Dynasty, there were many long-sleeve dance images in paintings and sculptures. To the Sui and Tang Dynasties, the absorption of India, Central Asia and other foreign forms, ribbon dance became more popular. For example, Amitayurdhyana Sutra illustration on the north wall of Cave 172 dated to the high Tang Dynasty, two people dancing with long ribbons, one showing the front, the other the back, both holding ribbons with both hands. The long ribbons bypass the dancers' back and fall from both arms, with the dance movements, hand-held ribbons up and down into loops, so that ribbons also have beautiful curves, which enhanced the performance of the dance. The same two-person dance also appeared in the Amitayurdhyana Sutra illustration on the east wall of Cave 148 (Fig. 6), where one person shows the front and one person shows the back, and the ribbons held by the dancers formed into circles, which the dance movement is obviously very rapid. In Cave 341, the two-person ribbon dance shows the dynamic of the ribbons rotating with the body, indicating that the dancers themselves are spinning. In addition to Cave 201 and Cave 112, the ribbon dance is a single dance, the movement is more leisurely and freely. The ribbon dance is the dance image which appears frequently in Dunhuang mural, dancers hold long ribbons to dance, from the ribbons revolving and fluttering image, we could feel the dance movements.

There are also many dancers dancing with waist drums, which is also a dance form influenced by the Western Regions. The dance in the Amitayurdhyana Sutra illustration in Cave 25 of Yulin Grottoes is typical. The dancers stand on one foot, wear long ribbons on their shoulders, look focused, and open their hands, as if they are trying to beat the waist drum (Fig. 7). The similar images can also be seen in Cave 108 and 156 of Mogao Grottoes. Waist drum dance is to beat a drum while dancing, we can imagine this is a bright rhythm, exciting dance art. In addition to waist drum dance, pipa dance is the image of dancers dancing while playing pipa. Pipa dance includes those who dance with pipa held in front and those who dance with pipa on their back. Playing pipa on back may have been a special skill of the dance at that time. Dancers raised their left hand and put pipa behind their back, while their right hand pluck strings inverted. There are images of back-playing pipa dance in Cave 172, 112 and 156, indicating that this style was popular in dance at that time. The back-playing pipa dance in Cave 112 is particularly typical (Fig. 8). The dancer is half naked, wearing a necklace and a brocade skirt, standing on the left foot and lifting the right leg high. The left hand raises the pipa to the back of head, and the right hand raises back to play. The body weight tilts to the right, and the ribbons rolled on both sides strengthen the dancer's shapes. Today, this back-playing pipa dancer image has become a representative dance image in Dunhuang murals. In many two-person dance scenes, either drum dance or pipa dance, or ribbon dance and pipa dance, all of them showing the wonderful and rich dance art of the Tang Dynasty. These different dance images are an inexhaustible source for today's dance art creation.

In Dunhuang murals, music and dance are mostly painted in sutra illustrations, which show the music and dance scenes of the Buddhist world. Among them, there are some imaginative elements, but some are real music and dance scenes of the secular world. These scenes are good materials to study. In the sutra illustration

Fig. 6 Dancers in Cave 148 of Mogao Grottoes (Copied by Wan Gengyu)

on the north wall of Cave 23 dated to the high Tang Dynasty, there is a scene of dance and music offering to stupa. In front of the square stupa, a group of musicians playing music on the square carpet, one dances in front of the stupa, and another person kneels piously in front of the stupa to worship. This is the illustration of worship Buddha scene by music and dance at that time. In the era when Buddhist belief infiltrates into daily life, music and dance activities were also inseparable from Buddhism. When expressing secular life scenes in sutra illustrations and story paintings, they often draw some independent music and dance scenes. For example, in the wedding picture of Maitreya Sutra illustration in Cave 445 dated to the high Tang Dynasty, in the banquet part, dancer in red waved her sleeves (Fig. 9), and several people nearby playing musical instruments, which truly reproduced the music and dance scene in ordinary people lives. On the south wall of Cave 85 dated to the late Tang Dynasty, Ulambana Sutra of Mahavaipulya Buddha illustration has the lost and blind prince playing zither under a tree, while the princess listening, this painting is quite vivid.

Fig. 7 Waist drum dance on the south wall of Cave 25 in Yulin Grottoes dated to the middle Tang Dynasty

Fig. 8 Music and dance scene on the south wall of Cave 112 at Mogao Grottoes dated to the middle Tang Dynasty

3. Clothes and popular colors

The open social environment of the Tang Dynasty contributed to the development of clothing fashion. The rich clothing images in Dunhuang murals have become the richest database for us to understand the clothing history of this era.

3.1 Clothes of emperors and ministers

Fig. 9 Music and dance in the wedding banquet in Cave 445 of Mogao Grottoes

When Dunhuang murals show secular figures listening Dharma or listening the debate of Bodhisattvas, they often have the images of emperors and common people to show the piety of secular figures at all levels to Buddha. For example, the emperor image in the Vimalakirti Sutra illustration of Cave 220 at Mogao Grottoes (Fig. 10), the emperor wears Mianliu on head, Mianfu on body, cyan Yi and red Chang, curved collar, white yarn Zhongdan, Dadai and Dashou. There are patterns of the so-called "Twelve Patterns" such as the sun, the moon, mountains and rivers on the clothes. These patterns can only appear on the clothes of the son of heaven since the pre-Qin Dynasty. Dashou have the painted image of dragon, all of which were in line with the clothing system at that time. From this, we can have a clear understanding of the emperor's Mian clothes explained in historical records. The clothes of the ministers around are mostly red yarn Danyi, white yarn Zhongdan, wearing Jinxian hat and Hu-head shoes, which were the court clothes of officials in the Tang Dynasty. Mink tails are also adorned on the hat of Ministers on both sides of the emperor. The Tang Dynasty classics recorded "when people serve as Shi Zhong, Zhong Shuling, left and right Sanqi Chang Shi, they will have mink tail on their hat", so we know the officials on both sides of the emperor who are Shizhong or other two kinds. This painting shows the emperor in a graceful manner, with ministers surrounded him, and can be compared with *the Thirteen Emperors Scroll* painted by Yan Liben (Fig. 11).

As a court painter, Yan Liben had the opportunity to see the emperor and officials by his own eyes. Of course, his paintings can truly reflect the dress characteristics of the emperor, ministers and other figures. Therefore, the consistency between Dunhuang murals and Yan Liben's works also shows the degree of authenticity and reliability. Cave 220 has the inscription of the 16th year of Zhenguan (642), which coincides with the period when Yan Liben lived. This means that after Yan Liben created the "emperor images" in Chang'an, it soon spread all over the country, and even Dunhuang in the northwest can get copies. Similar images of emperors and ministers can be seen in Cave 335 (686) of the early Tang Dynasty, Cave 103 and Cave 194 of the high Tang Dynasty. In Vimalakirti Sutra illustration, this become a fixed design, which had affected the murals of the late Tang and Five Dynasties.

3.2 Clothes of ordinary officials and common people

After the Wei and Jin Dynasties, under the influence of foreign clothes, men mostly wore long robes with narrow sleeves. However, until the Sui Dynasty, the robes were often short and called "big pleated clothes",

Fig. 10 The emperor profile on the east wall of Cave 220 at Mogao Grottoes dated to the high Tang Dynasty

Fig. 11 Yan Liben, *the Thirteen Emperors Scroll*, Museum of Fine Arts, Boston

which may be influenced by Kuxi of foreign clothes, with the characteristics of tight fitting, narrow sleeves and waist binding. The male donors of the Sui Dynasty in Cave 62 and Cave 281, we can see that their robes are slightly shorter and the boots are exposed. After the Tang Dynasty, the robe became longer, was ankle long and a horizontal line was added below the knees, which was called "Lanshan" or "Lanpao". When the images of ordinary men are shown in the murals of Cave 31, Cave 45, Cave 103 and Cave 217 dated to the high Tang Dynasty (Fig. 12), we can see that the robes as men's daily clothes were longer than those in the Sui Dynasty. The robes worn by the common people were often short, their boots were still exposed below the knees, and their robes were split on both sides so they can work more conveniently.

Officials wore Lanpao, usually have leather belts and leather boots. This was a more unified style after the reform of the clothing system in the Tang Dynasty. The formal shoes worn by ancient emperors and nobles were called "Lv" or "Xi", such as "square head shoes", "board head shoes", "cloud head shoes", etc. From the Han and Jin Dynasties tombs, many of those shoes have been unearthed. High-grade people generally have wooden soles and decorated shoe heads, which are bulky. The shoes were matched with the Han style wide robe

and large-sleeve clothes to show the noble demeanor, and they were often used in social occasion, maybe they are not convenient in real life. The boots were made by leather, similar to modern high leather boots, and they were originally used by nomads in the north. Foreign clothes were used in the Sui and Tang Dynasties, and leather boots were used by officials. In the tomb mural unearthed near Xi'an, the images of men in the Tang Dynasty wearing robes, leather belts, Fu hat and leather boots can also be seen.

Fig. 12 The image of men worshiping stupa in Cave 103 of Mogao Grottoes

The secular figures or donor figures in many cave sutra illustrations paintings, men are mostly wearing Fu hat, long robes with narrow sleeves and black leather boots, which were the common clothes of officials and scholars. After the high Tang Dynasty, men often wore long robes, such as the donor image Yue Tinggui, prefect of Jinchang County in Cave 130, wearing Fu hat and light cyan long robe (Fig. 13).

Fu hat, a popular hat for men since the Sui and Tang Dynasties, was originally to wrap hair by a piece of cloth and tie in the front and back. In the murals of the Sui Dynasty, there are two ends in the

Fig. 13 Donor image Yue Tinggui, prefect of Jinchang County, Cave 130, Mogao Grottoes (Copied by Duan Wenjie)

front and back of the cloth of the donor's hat (Cave 281), which is similar to the historical records that Emperor Wu of the Northern Zhou Dynasty "cut it into four ends". After the early Tang Dynasty, Ma Zhou suggested to Emperor Taizong of the Tang Dynasty that the form of Fu hat should be reformed and popularized throughout the country. From Dunhuang murals, it can be seen that the two ends hanging at the back of Fu hat in the early Tang Dynasty had become a fixed form, which was a common hat for men at that time. The form of Fu hat was gradually unified. The hat was divided into layers, low in the front and high in the back, with two ends hanging at the back. In the Tang Dynasty, a Jinzi (net) appeared, which was made by harder material, to fix the hair, and then tied it into a form of Fu hat with a piece of cloth. In this way, the Fu hat had a certain height. In the period of Wu Zetian, the Fu hat was high and leaning forward, which was called "all kinds of Wu family". At the time of Xuanzong, the dome shape was mostly used. In the late Tang Dynasty, it was completely made in the form of a hat, which was directly worn on the head and no longer have to wrap every day. The hanging ends at the back of the Fu hat were long or short, soft or hard, had a distinction of the times. In the early Tang Dynasty, the ends were generally soft, and some are long. In the late Tang Dynasty, there were wide and hard hanging ends. In the Five Dynasties, it developed into Fu hat with hard ends on both sides, which was more extended into Zhanjiao Fu hat in the Song Dynasty, we can see the complete track of the transformation of Fu hat form from Dunhuang murals.

3.3 Warrior clothes

From the images of the Maharāja-devas in the painted sculptures and murals, we can see the armor and equipment of warriors at that time, such as the Maharāja-devas in Cave 322 of the early Tang Dynasty, the helmet on the head, neck armor, arm armor, back armor, chest armor, body armor on the upper body, and battle skirt, bandage, and felt boots on the lower body. The Maharāja-deva of Cave 113 of the high Tang Dynasty wears neck armor, arm armor, beast head shoulder armor, leather over-sleeves, chest armor, armor belt, belly armor, battle skirt and leather boots. This set of armors are consistent with those records in *Tang Liu Dian · Military Equipment*, indicating that these Maharāja-devas' armor were shaped according to the military equipment at that time. Metal armor was mostly used in the Tang Dynasty, which was made of rectangular pieces of iron or copper. From the image of soldiers in the murals, we can see that the armor forms (such as Mingguang armor) at that time are consistent with historical records. In the picture of military training on the east wall of Cave 130, a man on the left is riding and shooting arrows, while several soldiers on the right are watching. They are all wearing metal armor made by square iron pieces, which are consistent with records in historical books (Fig. 14). In addition, the general's uniform includes war robe, split clothes and headband, etc. These are all appeared in the murals.

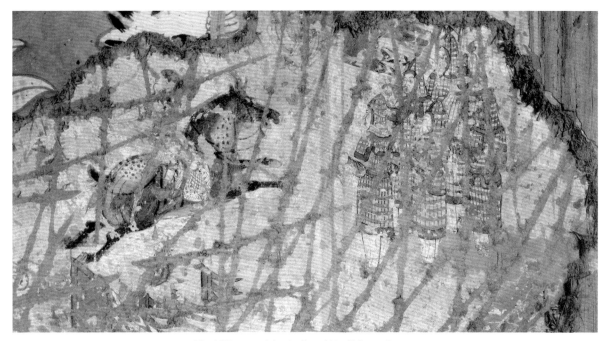

Fig. 14 Troop training in Cave 130 of Mogao Grottoes

3.4　Women's clothing

From the Six Dynasties to the Sui Dynasty, women mostly wore Ru skirt, the upper body was tight and narrow, and the lower skirt trained to the ground, reflecting the aesthetic trend of advocating thin and slim. Women in the Sui Dynasty also wore silk scarf, and the silk scarf was placed on shoulders, or on both sides of body, or around arms, with an elegant dynamic. After the Tang Dynasty, with the continuous expansion and opening to the outside world, women's fashion patterns were renovated and emerged one after another. The narrow shirt, small sleeves and half sleeves can be said to be the fashion of women in the early Tang Dynasty. The female donor on the east wall of Cave 329 kneels piously on the square felt and holds a lotus bud by both hands (Fig. 15). Her upper body is wearing a narrow shirt with small sleeves, the cuffs are very narrow, the round collar is open very low, and the skirt is tied high at the waist. There are two images of women on the east wall of Cave 217, a woman in the back can be seen wearing a modern short-sleeved shirt, which is half-sleeved coat (Fig. 16). The half-sleeved coat was originally worn by ordinary people, especially workers, and noble women in the Tang Dynasty often took it as a fashion. For example, in the mural of the tomb Princess Xincheng (663) unearthed in Li County near Chang'an, it can be seen that princesses and palace maids wear such half-sleeved coat, but noble women often wear half-sleeved coat over narrow-sleeved shirt, so that the skin of their arms will not be exposed.

Fig. 15 Female donors on the east wall of Cave 329 in Mogao Grottoes   Fig. 16 Women wearing half-sleeved coat on the east wall of Cave 217 in Mogao Grottoes (Copied by Duan Wenjie)

On formal occasions, noble women often wore large-sleeved Ru with wide cuffs, silk scarf, and long trained skirt. For example, the female donors in Cave 9 wear such clothes. Their clothes are very exquisite in texture and gorgeous in patterns, which show the characteristics of advanced dyeing technique. From the pattern, there are single and multicolor "wax dyeing" and "sewing and tying dyeing", showing the high achievements of dyeing technology in the Tang Dynasty.

In the donor image of the prefect's wife in Cave 130 (Fig. 17), the wife combed her hair into Paojia hair bun, wears Biluo flower shirt, a crimson flower half-sleeved coat, a red skirt, cloud head shoes and a white Luo flower shawl, with a graceful and rich demeanor. The thirteenth daughter who followed her wears a phoenix crown, hair accessories, flower makeup, half-sleeved shirt and long skirt, and small head shoes.

3.5　Women's hair style, face make-up and headdress

Women in the Tang Dynasty paid attention to many kinds of make-up, such as cover their face with vermilion powder, which was called "red make-up". There were mainly two kinds of red make-up in the early Tang Dynasty. One is "morning glow make-up" with vermilion on the forehead and upper eyelids, the Bodhisattvas in Cave 332 and Cave 57 make-up belong to this kind; the other is "black eyebrows and white face". According to the records of the Tang Dynasty, the court maids at that time "applied white powder to two cheeks, and call this tear make-up". Such makeup can be seen in painted donors and Bodhisattvas of Cave 329 and Cave 220, and the prefect's wife and daughters in Cave 130. Women in the Tang Dynasty loved to shape their eyebrows. In the early and high Tang Dynasty, it was popular to draw long eyebrows, called E'mei. The poet Zhang Hu of the Tang Dynasty wrote: "The lady of Guo State received the grace of the emperor and rode

Fig. 17 The prefect's wife in Cave 130 of Mogao Grottoes (Copied by Duan Wenjie)

into the palace in the morning. When she made up before she came, she disliked that makeup would stain her own beauty. So she just put on a light make-up and came to see the emperor." (*Ji Ling Tai Er Shou, the second*). After the high Tang Dynasty, short eyebrows were popular, such as the prefect's wife of Cave 130, all the women in the picture have short eyebrows (Fig. 18). This is the same as the characters in the picture of Group Portrait of Noble Women, which speculated as Zhou Fang's work.

The custom of women put flowers on face is very early. It is said that Shouyang Princess of Emperor Wu of the Southern Dynasty slept in the garden, and plum blossoms fell on her face. When she woke up, the plum blossom marks left on her face. Palace maids thought it was very beautiful and imitated. It was called plum blossom makeup. In the Tang Dynasty, there were many kinds of flower make-up on women's face. In the murals of the early Tang Dynasty, women already pasted flowers on their foreheads, just as the Tang poem said, "the green flower between the eyebrows is obvious". The prefect's wife in Cave 130 has five plum blossoms on her forehead and green flowers on her face. After the late Tang and Five Dynasties, it has been very popular for women to paste flowers or draw flowers on their face, this is very common in murals. In Maitreya Sutra illustration in Cave 25 of Yulin grottoes, it can be seen that women have red flowers on their forehead (Fig. 19). In the late Tang and Five Dynasties, this trend was even stronger. Because of the large size

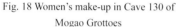

Fig. 18 Women's make-up in Cave 130 of
Mogao Grottoes

Fig. 19 Women's headdress and makeup in Cave
25 of Yulin Grottoes

of the donor figures in the murals of the Five Dynasties, they are particularly clear. For example, the female donor figures in Cave 98 and Cave 61, the upper half of their face are painted with rouge in the shape of a half moon and a flower on it. The cheeks are often painted with phoenix, butterfly or flower, which are extremely abundant.

The hair style of women in the Tang Dynasty was also very particular. In the early Tang Dynasty, there were many high hair bun, which can be divided into bar hair bun and half turned hair bun. After Kaiyuan Tianbao period, Paojia hair bun was popular, the hair on both sides of the face hang down to cover the face, and the upper hair was made into a certain shape upward, such as bar shape, flower shape, etc. Most women in the prefect's wife picture in Cave 130 of Mogao Grottoes are Paojia hair bun. They are not only common in Dunhuang murals, but also in paintings handed down from the Tang Dynasty, such as the painting of *Ladies Waving Fans* speculated by Zhou Fang.

In the early time of the Tang Dynasty, women often put flowers in their hair or didn't put any accessories. Then more and more decorations were put on women's head in the later time of the Tang Dynasty, especially in the late Tang and Five Dynasties. Hairpins, combs and other decorations on women's head became more and more popular. For example, the donors of Cave 9 in the late Tang Dynasty painted flowers on their foreheads, and their hair are filled with hairpins and flowers, up to more than ten pieces (Fig. 20). After the Five Dynasties, the decoration of women's head was more complex, and the forms of facial flowers were also rich and diverse. For example, the female donors of Cave 98 and 61 are the representatives. In addition to the hairpins, some women such as the Uyghur princess also wear the peach-shaped crown and phoenix crown of the Uyghu nationality, they also put Buyao as decoration which is incomparably gorgeous.

3.6 Popular colors in the Tang Dynasty

The culture of the Tang Dynasty was full of fashion, and the popular colors of the Tang Dynasty were constantly renovated with the development of society. The cave murals in the early Tang Dynasty have fresh

Fig. 20 Female donors of Cave 9 in Mogao Grottoes (Copied by Ouyang Lin and Shi Weixiang)

colors, just like the spring breeze and all things sprout. The light turquoise has become the most popular color in murals. Cave 220 of Mogao Grottoes, built in the 16th year of Zhenguan, maintains the original color. The mural color in this cave is mainly bright turquoise. In the Western Pure Land illustration, light green was used to show the pond, lotus are mostly depicted in blue, and there are many green trees, forming a very coordinated turquoise tone, showing a bright, clean and elegant atmosphere. The Bhaisajyaguru Sutra illustration on the north wall of the same cave shows the Pure Glass World of the Eastern Pure Land. The ground is mostly interwoven with ochre red and white flowers to show the precious stones such as glass and agate. The background color of the whole wall is still dominated by turquoise. The seven body Buddhas wear dark ochre or earth red kasaya, which coordinate with the green background and ochre red on the ground. It can be said that the mural color in Cave 220 is a light and elegant style by turquoise intertwined with ochre. As Cave 321 of Mogao Grottoes dated to the early Tang Dynasty, the Ten Wheels Sutra illustration on the south wall used the magnificent green landscape as the background, and the costumes of the characters are either green consistent with the background, or gorgeous earth red and vermilion, showing a bright and lively style. The north wall of the cave shows the Western Pure Land illustration. The main pictures are the pavilions and platforms, railings, bridges and other buildings in the green water pond. A considerable area in the upper part of the picture shows the scene of the sky, the dark blue as the sky color. There are treasure pavilions, treasure trees, self playing musical instruments and flying Apsaras in the sky, showing the clarity of the air. The mural on the west wall of the cave and the top of the niche also match the color of the north and south walls to show the flying Apsaras in the dark blue sky. The tone of this cave is dominated by the dark blue of the sky, combined with the green landscape and other scenery, which has a distinctive and elegant style. Another type of grottoes in the early Tang Dynasty is murals dominated by earth red, such as Cave 57, 322, 328, 329 and 335 in Mogao Grottoes. The murals in the west niche of Cave 329 are mainly in red, showing the stories of Buddha's life; the four slopes on the top of the cave covered by Thousand Buddhas in earth red background; the whole sutra illustrations on the north and south walls take earth red ochre as the main tone to show the temple buildings of the Buddhist world, supplemented by dark blue and malachite green to show the sky and pond. The whole cave presents a warm and high-spirited atmosphere.

In the high Tang Dynasty, the colors of cave murals were richer. In the high Tang dynasty caves represented by Cave 23, 217 and 148, the turquoise was still the most popular color. But at this time, the malachite green gradually used as the main color, which is more calm and thick than the turquoise in the early Tang Dynasty. If the turquoise in the early Tang Dynasty reflects the spring, the malachite green at this time is the dense forest in midsummer. While malachite green was popular, there was also a change in ochre red tone, that is, the red parts are more distinct and prominent, and coexist with the same deep blue, green and some black colors in paintings, making the picture more rich and profound. The mural colors of Cave 320, Cave 171 and Cave 172 also reflect the tendency of richness, beauty and brilliance in their maturity and calmness.

Some murals with simple colors appeared in the end of the high Tang Dynasty, and became popular after the middle Tang Dynasty. The number of colors in these caves are reduced, and the colors are light and elegant. Malachite green dominates all the murals, forming another appearance of the caves in the late Tang Dynasty. Represented by Cave 159 and 196 of Mogao Grottoes, malachite green and azurite were used lightly, and white or light yellow were often used, which make the overall style of the caves light and elegant.

The secular figures clothes in the murals of the Tang Dynasty also provide us the popular colors of people clothes at that time. The tendency embodied in the clothes of these figures is also roughly consistent with the general tone of the murals. Men's clothes are mostly round collar robes with little change. They are mostly red, ochre and black, as well as white and blue, generally they are mainly dark ochre red. There are many changes in women's clothes. In the early Tang Dynasty, pink, light yellow and light blue clothes can be seen. After the high Tang Dynasty, clothes became brilliant and colorful, with a wide range of flower patterns and rich changes in layers, showing the highly developed technology of textile and dyeing in the Tang Dynasty. At the same time, in the mural figure decoration of the Tang Dynasty, gemstone, jade, glass, agate and other materials also abundant, adding infinite and rich interest to the popular colors of this era.

In short, the popular colors of the Tang Dynasty were very rich, but there were some basic colors that were throughout the Tang Dynasty, such as turquoise, red ochre, black and brown. The color of the early Tang Dynasty was bright and gorgeous, the color of the high Tang Dynasty was calm and thick, and the color of the middle and late Tang Dynasty was graceful and simple. The color application in clothing, architectural decoration and utensils were also different, but under the general trend of popular color, all aspects of color will change accordingly.

4. Foreign fashion and international exchange

The Tang Dynasty was a very open era, it widely accepted foreign cultures, enriching itself and strengthening China. In the Sui and Tang Dynasties, the cultures of the Western Regions flowed into China, and there were foreign clothes, foreign tent, foreign bed, foreign seat, foreign food, foreign Konghou, foreign flute, foreign dance and so on. As Lu Xun said, "the Tang Dynasty was full of foreign atmosphere". This "foreign trend" phenomenon is also reflected in Dunhuang art. Dunhuang, located in the hub of the Silk Road, had more frequent contacts between Chinese and foreign envoys and businessmen than mainland cities. In addition, people from many countries and nationalities often lived here. The exchanges between ancient China and Western countries have left rich image materials.

According to Chinese historical records, grapes were brought from Dayuan after Zhang Qian's mission to the Western Regions. *The Historical Records* also recorded that Dayuan people use grapes to make wine, and people like to drink wine. According to relevant historical research, grapes are native to the region between the Caspian Sea and the Black Sea in Asia Minor and its south bank. Persia (now Iran) and Egypt are the first two ancient countries to cultivate grapes and make wine by grapes. After the Han Dynasty, grapes and wine were introduced into China. According to the historical records from the Han Dynasty to the Tang Dynasty, the origin of grapes gradually developed from western Persia to the east. Up to the Tang Dynasty, grapes were planted and brewed in Qiuci (now Kuqa), Qiemo, Yutian (now Khotan), Gaochang (now Turpan), Yizhou (now Hami) and other places in the Western Regions. *Tang Hui Yao* and other documents also recorded that Emperor Taizong of the Tang Dynasty took Gaochang and introduced grape planting and wine brewing technology to China. Since then, wine can also be made in Han nationality areas. The wine producing areas of the Tang Dynasty mainly included Xizhou (now Turpan), Liangzhou (now Wuwei, Gansu), and Bingzhou (now Taiyuan, Shanxi). Nevertheless, grapes and wine were still very precious things in the Tang Dynasty. Bao Fang, a Jinshi at the end of Tianbao, wrote in his poem "The flying horse often holds alfalfa flowers in mouth and foreign people pay tribute by wine", indicating that wine was a very precious thing as a tribute from Western countries at that time. In the murals and handicrafts of the Tang Dynasty, we can see many decorative patterns of grapes (Fig. 21), indicating people's love for grapes at that time.

Fig. 21 Grape pomegranate pattern caisson ceiling in Cave 209 of Mogao Grottoes (Copied by Duan Wenjie)

Lion is native to Africa and West Asia. Some scholars speculate from the pronunciation of "Lion" that they are from Persia. Chinese history recorded that as early as the Western Han Dynasty, lions had been imported into China from Western countries. Due to the ferocity of the lion, the image of lion was often used as a tomb beast after the

Eastern Han Dynasty, which was called exorcism. After the Tang Dynasty, there were more records of Western countries offering lions. For example, *Tang Hui Yao* recorded that in July of the ninth year of Zhenguan (635), the state of Kang "offered lions, and Taizong praised they came far, so ordered Yu Shinan of Mishujian to write records for them." In addition, there are records of lions offered by Miguo, Tuhuoluoguo and other countries from the Western Regions in the Tang Dynasty. Lion has a sacred position in Buddhism. For example, the mount of Manjusri Bodhisattva is also a lion. Therefore, Chinese people often regard lion as sacred animal. So artists at that time could see the actual lion, and the lion in paintings should reflect the reality. In Dunhuang murals we can see not only Manjusri riding a lion, but also Kunlun Slave leading the lion (Fig. 22). Kunlun Slave has brown complexion and curly hair, which may have come from ancient Southeast Asia. In the mural, the white elephant ride by Samantabhadra Bodhisattva is also led by Kunlun Slave. This is also reasonable for these animals from foreign countries to be led by foreign people.

Elephants were also foreign animals in the Tang Dynasty. Although elephants once existed in northern China in the bronze age, we can see many decorative images of elephants in the bronzes of the Shang and Zhou Dynasties, with the climate change,

Fig. 22 Lions and Kunlun Slave in Cave 25 of Yulin Grottoes

Fig. 23 Foreign people lead elephant in Cave 103 of Mogao Grottoes

elephants finally disappeared in the north. After Buddhism came to China, there are a lot of elephant-related Buddhist stories, because India is also a country has elephant. Therefore, the image of elephant appears from time to time in Buddhist Art. In the Tang Dynasty, foreign elephants also came to China in an era of unprecedented exchanges between China and foreign countries. According to the historical records of the Tang Dynasty, Linyi, Zhenla and other countries in the south once paid elephants as tribute in the Tang Dynasty. Of course, when these elephants arrived in China, they became rare animals. They were usually locked up in forbidden gardens. When there are celebrations, they are led out for people to watch and increase the lively atmosphere. Nevertheless, elephant closely related to Buddhism always has been seen by Chinese people. Therefore, the elephant painted in Buddhist Art has authenticity (Fig. 23).

In the Sui and Tang Dynasties, because the Silk Road was unobstructed, there were many exchanges with foreign countries.

Under the influence of foreign countries, the so-called "foreign clothes" was also widely popular in the Han culture area. For example, the "foreign hat" that noble men and women liked to wear when traveling was a imitation to the clothes of Turks and East Iranians. In the early Tang Dynasty, women often wore "Mi Li" which is a combination of foreign hat and veil to cover their head and face. "Mi Li" was often matched with coat similar to cloak. Such clothes not only protect noble women away from wind when they go out, but also hide them to be seen away from outsiders. Later, it was shortened and called the veil hat. The veil around the hat can prevent wind and sand. This can be said a reformed form of foreign hat. In Cave 217 of Mogao Grottoes, a figure wearing a red cloak and a veil hat riding on horseback, moving in the mountains (Fig. 24).

Fig. 24 A person wearing veil hat on the south wall of Cave 217 in Mogao Grottoes

Lapel coat is a popular style of foreign clothes in the Western Regions, which was popular in western China and even Central Asia. In the existing murals in Central Asia and Kucha, you can see Sogdian and Kucha people in Lapel coat. The fabric of this coat is hard, which may be made of leather or thick wool fabric. Such clothes became a fashion among the Han nationality in the Central Plains during the early Tang Dynasty. Nobles can be seen wearing such clothes in tomb murals in Chang'an area. For example, there are many figures wearing foreign clothes in the murals of Wei Hao's tomb (708) unearthed in Shaanxi Province, and some women wear men's foreign clothes, which was also a more fashionable way of wearing at that time. In Dunhuang murals, you can also see the images of

Fig. 25 A person wearing foreign clothes on the east wall of Cave 217 in Mogao Grottoes (Copied by Li Qiqiong)

men wearing foreign clothes, such as the figures on the east wall of Cave 217 (Fig. 25), the main figures in the wedding of Cave 445, etc.

Dunhuang had frequent contacts with ethnic minorities in the Western Regions. There were many Persians, Kucha people, Uighurs and Tubo people lived in Dunhuang in the Tang Dynasty. It is not surprising that the Han nationality wore foreign clothes and hats. A large number of foreign figures or ethnic figures in the Western Regions in the murals also provide rich image materials for us to understand the history of the foreign and Han blending at that time (Fig. 26).

Fig. 26 Foreign figures in Cave 103 of Mogao Grottoes

From the Northern and Southern Dynasties to the Sui and Tang Dynasties, a large number of Persian handicrafts were imported to China through the Silk Road. For example, hunting scenes were the most popular theme in ancient Persia. Many silver plates have hunting images were found in northern China. The Sassanian Dynasty of Persia (3rd-5th century) was contemporaneous of the golden era of the Roman Empire. Roman culture in Europe, Persian culture in West Asia and Chinese culture collided in a wide area from Central Asia to West Asia, leaving civilizations blending footprints everywhere.

From the Sui Dynasty to the early Tang Dynasty, there are a large number of Persian patterns on Dunhuang painted sculptures and in murals of Bodhisattva's clothing, such as Sassanian roundels with bird pattern, Sassanian roundels with beast pattern, Sassanian roundels with flying horse pattern and so on. Until the early Tang Dynasty, the Persian style beads pattern was still very popular. Among the textiles unearthed in Xinjiang, there are also beads patterns. At first, such linked beads pattern came from Persian fabrics, but after the Tang Dynasty, China could also make brocade according to Persian technology. Therefore, the brocade of the Tang Dynasty unearthed along the Silk Road, some of their pattern style and production method are completely Persian, but weaved with Chinese characters. The introduction of Sassanian roundels pattern had a far-reaching impact on the decorative patterns of the Tang Dynasty. The love for the beads pattern gradually became a part of the Chinese traditional decoration.

In addition, grape pattern and pomegranate pattern derived from Central Asia were also widely used in Dunhuang Grottoes, and constantly been modified. They were combined with scrolling vine pattern and became into very rich decorative patterns, which shine brightly in the murals of the early and high Tang Dynasty (Fig. 27 and Fig. 28).

Glass and coloured glaze were once very precious things. When they were first introduced into China, only emperors and some nobles could have them. Therefore, in Dunhuang murals of the Sui and Tang Dynasties, Bodhisattvas are often painted as holding glassware to show their dignity. Glassware has bowl shape inlaid with jewels and cup shape, showing the manufacturing technology at that time.

Incense burners are also indispensable for Buddhist worship. They were often designed based on foreign styles. The incense burners in the Tang Dynasty were exquisite in workmanship and various in styles. The gold-plated silver censer unearthed from the underground palace of Famen Temple has a six-leg base, a lotus-bud-shaped censer cover, and pendants on the base and the edge of the container. Similar incense burners are often seen in Dunhuang murals. The incense burner popular in the murals of the late Tang Dynasty has a lotus-shaped base at the lower part, an octagonal or hexagonal opening at the upper part, a circular hollow cover on the top, and a jewel is often decorated in the center.

Fig. 27 Pomegranate scrolling
vine pattern in Cave 23 of
Mogao Grottoes

Fig. 28 Grape pomegranate scrolling vine pattern head light in Cave 444 of
Mogao Grottoes

5. Summary

The political openness and highly developed economy of the Tang Dynasty, especially the prosperity of the Silk Road, greatly promoted the exchanges between China and foreign countries. The exchange and mutual learning between foreign culture and Chinese traditional culture provided environment for the prosperity of culture and art in the Tang Dynasty. From the perspective of fashion, the art of the Tang Dynasty is full of creativity. Here in Dunhuang mural alone, is already endless, constantly innovative, dazzling treasure. Limited to the space, this article only makes a discussion on the above several aspects, in a narrow view only.

敦煌莫高窟盛唐第45窟彩塑菩萨像服饰分析

齐庆媛

敦煌莫高窟至今尚存十六国至西夏彩塑两千余身，不仅对了解完整的石窟艺术发挥着主体性作用，还为研究中国古代服饰文化提供了宝贵资料。

本文选取莫高窟盛唐最具代表性洞窟——第45窟的两尊彩塑菩萨像为研究对象，着眼于发髻与宝冠、装饰物、服装三方面，在联系莫高窟其他洞窟相关实例的同时，尽可能与中原北方地区唐代石刻菩萨像进行比较分析，期望以点带面梳理莫高窟盛唐彩塑菩萨像服饰的特征，并阐明与世俗服饰之间的内在联系。

一、基本情况

莫高窟第45窟盛唐彩塑一铺位于西龛佛坛，现存七身，按照对称原则配置，中间为结跏趺坐佛像，其两侧由内而外依次为弟子、菩萨、天王立像，加之龛外两侧已毁的力士像，构成一铺九身式造像组合，与洛阳龙门石窟奉先寺上元二年（675年）完工的大卢舍那佛像龛布局一致。两尊菩萨像造型比较统一（图1、图2），高185厘米，束高发髻，佩戴胸饰璎珞、臂钏与手镯，上身斜披络腋，下身着长短套裙，颔首低眉，头部、上身、胯部扭倾形成三曲身段，娉婷婀娜、温婉健美，充溢着生命活力。这种风格的菩萨像在唐武周时期至开元年间（690—741年）盛行一时。就服饰的整体搭配而言（图3），莫高窟第45窟菩萨像除了没有表现披帛外，其他均与关中地区的诸多实例高度接近，如长安光宅寺七宝台武周时期石刻造像、耀县❶药王山开元年间摩崖造像（图4）。如果具体到服饰的细部造型，莫高窟第45窟菩萨像并非直接模仿关中地区样式，而是受到各种因素影响，那么应将其置入更广阔的视野来探究。

图1　敦煌莫高窟第45窟西龛南侧　盛唐彩塑菩萨像［出处：段文杰主编《敦煌石窟艺术·莫高窟第四五窟附第四六窟（盛唐）》，南京：江苏美术出版社，1993年，图31］

图2　敦煌莫高窟第45窟西龛北侧　盛唐彩塑菩萨像［出处：段文杰主编《敦煌石窟艺术·莫高窟第四五窟附第四六窟（盛唐）》，南京：江苏美术出版社，1993年，图25］

❶ 今陕西省铜川市耀州区。

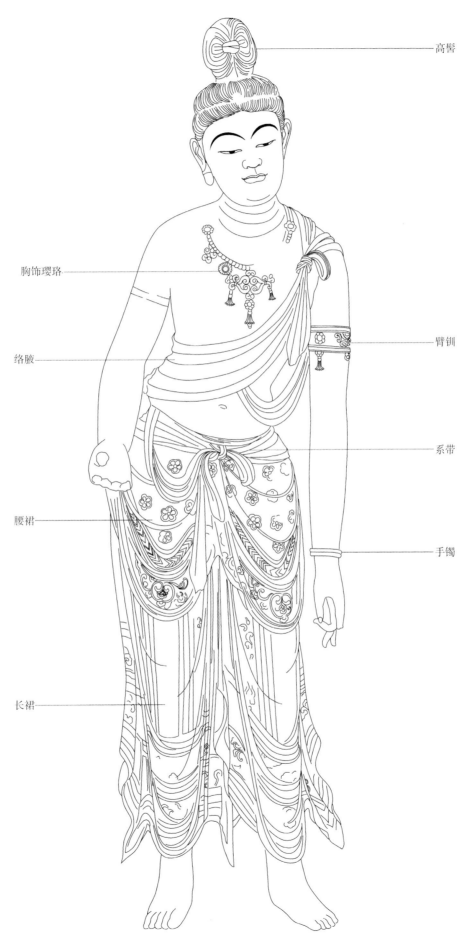

高髻

胸饰璎珞

臂钏

络腋

系带

腰裙

手镯

长裙

图3　敦煌莫高窟第45窟西龛北侧　盛唐彩塑菩萨像服饰名称示意图（齐庆媛绘）

图4　耀县药王山摩崖第9龛　唐开元十一年（723年）
双观世音菩萨像（齐庆媛摄）

二、发髻与宝冠

莫高窟第45窟两尊彩塑菩萨像一缕缕头发规整地梳至头顶形成高髻，根据头发所在位置可以分为额发、中发和顶发三部分，另外还有下垂至耳的鬓发（图5）。这种三段式高发髻为唐代菩萨像的典型特征（图6）[❶]。头顶高髻在中间用束带系后分至左右两侧呈叠环状，更成为初盛唐时期菩萨像最为流行的发髻样式，在石窟、摩崖造像及地面寺院出土造像中屡见不鲜。两尊菩萨像鬓发作绕耳表现，保存较好的南侧菩萨像左耳处鬓发分为两缕，似乎承袭了彬县[❷]大佛寺石窟大佛洞西壁初唐观世音与大势至菩萨像鬓发形式（图7），这种表现在唐代仅仅属于个别情况，至宋代及其以后逐渐流行。

束带　摩尼宝珠　联珠　冠箍　团花圆形宝珠　鬓发

顶发　叠环状　中发　宝冠　额发　鬓发绕耳

图5　莫高窟第45窟西龛南侧　盛唐彩塑菩萨像发髻与宝冠名称示意图（齐庆媛绘）

---

❶ 中国国家博物馆藏。
❷ 今陕西省彬州市。

唐代女子梳高髻蔚然成风，当时所流行的半翻髻、回鹘髻、惊鹄髻、椎髻、螺髻等均属于高髻。通过唐诗记载可以进一步了解高髻的受欢迎程度。诸如，白居易《江南喜逢萧九彻因话长安旧游戏赠五十韵》："时世高梳髻，风流澹作妆。"[1]元稹《李娃行》："髻鬟峨峨高一尺，门前立地看春风。"[2]虽然唐代菩萨像高髻的具体形态难以与世俗发髻完全对应，却反映了当时的审美情趣与时代风尚。需提及的是，陕西礼泉唐乾封二年（667年）韦贵妃墓女乐伎[3]，鬟发下垂拂过耳畔，为研究初盛唐少数菩萨像鬟发绕耳提供了启示。

莫高窟第45窟西龛南侧菩萨像戴宝冠，冠箍装点两个团花圆形宝珠，上缘饰一周联珠，其上左右侧各安置一颗摩尼宝珠（同图5）。冠箍与摩尼宝珠相结合的形式，已经体现在莫高窟隋代洞窟彩塑与壁画菩萨像宝冠上，初盛唐时期延续发展。然而，绝大多数宝冠正中还有一颗摩尼宝珠，与两侧饰物构成典型的三珠冠。再从发髻与宝冠两方面因素考虑，相似造型与组合形式可以对比西安东关景龙池出土的唐代观世音菩萨像（图8）[4]，但是其宝冠正中设置了化佛以表明尊格。如同莫高窟第45窟西龛南侧菩萨像那般，仅在宝冠两侧表现饰物的情况罕见。所以不排除其宝冠正中原初设计了摩尼宝珠或表明尊格属性装饰的可能性，遗憾的是由于时间久远未能保存完整。西龛北侧菩萨像额发与中发之间存在一周的磨损痕迹（图9），推测其原初也应戴宝冠。

图6　太原天龙山石窟第18窟　唐代菩萨头像（齐庆媛摄）

三、装饰物

莫高窟第45窟两尊彩塑菩萨像佩胸饰璎珞，戴臂钏和手镯，比较接近中原北方地区诸多石刻造像，集中反映了唐代菩萨像装饰简洁的特征。

两尊菩萨像的胸饰璎珞是由项圈和联珠垂挂、串联各种饰物而成的（图10、图11），可以分为主要饰物和次要饰物详细分析。

图7　彬县大佛寺石窟大佛洞西壁　初唐观世音菩萨像（齐庆媛摄）

---

[1]（清）彭定求等编，中华书局编辑部点校《全唐诗》（增订本）卷462，北京：中华书局，1999年，第5282页。
[2]（唐）元稹《元稹集》，北京：中华书局，1982年，第698页。
[3]徐光冀主编《中国出土壁画全集6·陕西（上）》，北京：科学出版社，2011年，图197、图198。
[4]西安碑林博物馆藏。

图8　西安东关景龙池出土唐代观世音菩萨像局部
（齐庆媛摄）

图9　敦煌莫高窟第45窟西龛北侧　盛唐彩塑菩萨像局部〔出处：段文杰主编《敦煌石窟艺术·莫高窟第四五窟附第四六窟（盛唐）》，南京：江苏美术出版社，1993年，图26〕

图10　敦煌莫高窟第45窟西龛南侧　盛唐彩塑菩萨像胸饰璎珞复原线描（齐庆媛绘）

图11　敦煌莫高窟第45窟西龛北侧　盛唐彩塑菩萨像胸饰璎珞复原线描（齐庆媛绘）

南侧菩萨像胸饰璎珞主要饰物为一朵硕大团花，团花由内外两部分构成，内为小联珠环绕大圆形宝珠，外为6个涡卷C形背向相连（最上方还有一个C形仅表现了部分，似乎是受限制形成），其创意和造型类似洛阳龙门石窟擂鼓台中洞武周时期宝冠佛像胸饰璎珞的团花，只是后者团花6个涡卷C形为正向相连（图12），形式略有不同。一个个涡卷C形背向或正向相连围绕一周，构成唐代典型纹样——宝相花的重要组成部分（图13、图14），涡卷C形尤以6个或8个居多，由此可知唐代菩萨像或宝冠佛像涡卷C形团花装饰之来源。

图12　洛阳龙门石窟擂鼓台中洞　武周时期宝冠佛像胸饰璎珞线描（齐庆媛绘）

图13　敦煌莫高窟第334窟　初唐窟顶藻井局部［出处：敦煌文物研究所编《中国石窟　敦煌莫高窟（三）》，北京：文物出版社，1987年，图83］

图14　敦煌莫高窟第320窟　盛唐窟顶藻井［出处：敦煌文物研究所编《中国石窟　敦煌莫高窟（四）》，北京：文物出版社，1987年，图8］

044

图15 敦煌莫高窟第328窟西龛南侧 盛唐彩
塑菩萨像局部［出处：敦煌研究院编《中国美
术全集·雕塑编7敦煌彩塑》，上海：上海人民
美术出版社，1989年，图126］

图16 敦煌莫高窟第45窟西龛北侧 盛唐彩塑
菩萨像臂钏线描（齐庆媛绘）

图17 敦煌莫高窟第45窟西龛北侧 盛唐彩塑
菩萨像手镯线描（齐庆媛绘）

北侧菩萨像胸饰璎珞主要饰物为一对左右对称的卷草，卷草呈双茎桃形头部向心颈部束茎式，附加内卷小叶芽。类似的卷草饰物，频繁见于第386窟、205窟、320窟、319窟、328窟等窟菩萨像的通身璎珞、胸饰璎珞与臂钏（图15），成为莫高窟初盛唐彩塑菩萨像饰物的显著特征。卷草饰物的创意与外形，酷似南北朝隋代双茎桃形头部向心颈部束茎式忍冬纹❶，应由此种形式的忍冬纹发展而来。

两尊菩萨像胸饰璎珞次要饰物小团花、小联珠环绕大宝珠与流苏，北朝晚期以来广泛应用于菩萨像装饰。初盛唐时期，胸饰璎珞点缀流苏的情况，也常见于洛阳龙门石窟与太原天龙山石窟造像，可以与莫高窟第45窟菩萨像做对照。

两尊菩萨像的臂钏大体一致（图16），两行弦纹之间装饰小团花与涡卷C形背向相连的大团花，下缘垂流苏，表明臂钏与胸饰璎珞基于相同设计构思，莫高窟盛唐彩塑菩萨像饰物普遍存在这种特征。两尊菩萨像的手镯雷同（图17），为双股圆环式，这是唐代菩萨像极为盛行的手镯样式，简洁的造型与浙江长兴下莘桥唐代窖藏出土双股银钏相似❷。

有必要提及的是，莫高窟第205窟中心佛坛北侧盛唐彩塑菩萨像的胸饰与臂钏（图18），将涡卷C形背向相连的大团花与左右对称的卷草组合在一起，仿佛是从第45窟两尊菩萨像主要饰物的融合中获得新生。

四、服装

1. 服装形制

莫高窟第45窟两尊彩塑菩萨像皆上身斜披络腋，下身穿长短重裙，这是唐代菩萨像最为流行的服装形制，风行大江南北。

络腋表现为一块长条状丝帛，斜挎背后，一段从右腹斜向上，另一段自左肩下垂，两段在左胸上部系结，形成自左肩斜至右腹的形式❸。两尊菩萨像络腋同中存异，系结后末端的塑造有所区别，显得灵活多变。南侧菩萨像络腋较长一段下垂至腹部，末端塞入腹前丝帛内；另一段绕左肩搭在身后（同图1）。北侧菩萨像络腋较长一段自左肩下垂至身后再绕至体前，末端塞入丝帛内；另一段在左胸部系单花结后

❶ 李秋红《南北朝隋代双茎桃形忍冬纹分析》，中国古迹遗址保护协会石窟专业委员会、龙门石窟研究院编《石窟寺研究》（第8辑），北京：科学出版社，2018年，第209–254页。
❷ 浙江省博物馆编、石超主编《错彩镂金：浙江出土金银器》，杭州：浙江人民美术出版社，2016年，第29页上图。
❸ 以物象自身为基准判断左右方向，全文皆同。

自然下垂（同图2、图3）。纵观唐代菩萨像络腋，自左肩斜至右腹者比比皆是，自右肩斜至左腹者也不在少数，前者明显受到印度笈多朝菩萨像服装的影响（图19）❶，后者很可能在中国本土形成。莫高窟唐代菩萨像的络腋也包含这两种形式，与中原北方地区石刻造像同步发展。然而绝大多数情况下，两种络腋末端在左肩胸或右肩胸处敷搭缠绕，系结的做法零星见于太原天龙山石窟造像，却在莫高窟备受欢迎，诸如初唐第322窟、332窟与盛唐第66窟、79窟、264窟、320窟、444窟、458窟的众多实例，因此可以将络腋系结视为莫高窟初盛唐彩塑菩萨像服装的地域性特征。

图18　敦煌莫高窟第205窟中心佛坛北侧　盛唐彩塑菩萨像局部［出处：敦煌文物研究所编《中国石窟　敦煌莫高窟（三）》，北京：文物出版社，1987年，图123］

图19　北方邦鹿野苑遗址出土　5世纪弥勒菩萨像局部（齐庆媛摄）

　　两尊菩萨像下身内着及踝长裙，外裹短围裙，然后用条带将内外两重裙装系结，条带大同小异，饶有趣味。南侧菩萨像条带两段自结节处下垂形成两个"U"形，末端塞入胯部裙内（同图1）。北侧菩萨像条带一段形成"U"形，末端塞入胯部裙内，另一段系单花结后自然下垂（同图2、图3）。内长外短两重裙装是唐代菩萨像下裳的典型形制。裙外再系带的做法，在中原北方地区流行于武周至玄宗开元前后，主要分布在关中、洛阳周围、太原和青州等地。莫高窟第45窟两尊菩萨像裙装及系带的表现，尤其接近荥阳大海寺遗址出土的唐武周时期石刻造像（图20）❷。

❶ 鹿野苑考古博物馆藏。
❷ 荥阳大海寺遗址出土石刻菩萨群像，时间跨度大约涵盖武周、盛唐、中唐和晚唐四个阶段。李静杰《关于荥阳大海寺遗址出土唐代石刻菩萨像的再认识——读〈河南佛教石刻造像〉收获》，《中原文物》2010年第4期。

2. 服装色彩与纹样

两尊菩萨像服装以红、绿为主，搭配黑、蓝、金黄诸色，形成斑斓多姿、富丽堂皇的视觉效果，展现了盛唐的审美趣味。络腋、长裙底色及系带呈石榴花般的朱红色，沉稳艳丽。短裙底色接近翡翠绿，鲜亮纯净。朱红翠绿互为对比色，色彩饱和度高，对比强烈，耀眼夺目。红绿两色在唐代女子服装中占据主流地位，组合搭配的情况时有所见，不但直观体现在唐墓出土的纺织品、绘画与陶俑人物的着装上（图21）❶，而且可以根据大量唐诗作品的记载客观论证。诸如武则天《如意娘》："不信比来长下泪，开箱验取石榴裙。"❷《武后时童谣》："红绿复裙长，千里万里犹香。"❸毛文锡《赞浦子》："懒结芙蓉带，慵拖翡翠裙。"❹元稹《晚宴湘亭》："舞旋红裙急，歌垂碧袖长。"❺这样看来，莫高窟第45窟菩萨像服装色彩

图20　荥阳大海寺遗址出土　唐代光相菩萨像（出处：河南博物院编、王景荃主编《河南佛教石刻造像》，郑州：大象出版社，2009年，唐代造像之图版22）

图21　西安市王家坟村出土　唐三彩女子坐俑（出处：大阪市立美术馆《〈大唐王朝女性の美〉展》，中日新闻社，2004年，图39）

---

❶ 陕西历史博物馆藏。

❷《全唐诗》（增订本）卷5，第60页。

❸《全唐诗》（增订本）卷878，第10015页。

❹《全唐诗》（增订本）卷893，第10155页。

❺《全唐诗》（增订本）卷409，第4559页。

合乎唐代红绿相衬的审美风尚，再装饰黑、蓝、金黄等多种色彩的纹样，显得格外绮丽华美，较世俗女子服装有过之而无不及。

两尊菩萨像裙装描绘的纹样（图22）可以分为三种。

其一，五瓣花朵纹，清新朴素、简约灵动，分散点缀于短裙。花朵纹是唐朝流行的纺织品纹样，有四瓣、五瓣、六瓣多种形式，不但是世俗人物服装的常用图案（同图21），也应用于出家僧人的服装，如莫高窟盛唐第199窟西龛壁画僧人、第320窟西龛内北侧彩塑僧人，皆身披红绿两色相间的五瓣花朵纹袈裟。初盛唐时期，莫高窟众多壁画和彩塑菩萨像的披帛、络腋、围裙与长裙等装点各种各样的小花朵纹，折射出当时人们对这类纹样的由衷喜爱。

其二，卷草纹，左右起伏的纤细波状枝叶在相交处滋长曲状花瓣，装饰在短裙下缘及长裙边缘。卷草纹在唐代迎来大发展契机，其形式之多样、应用之广泛，达到登

图22　敦煌莫高窟第45窟　盛唐彩塑菩萨像裙饰织锦图案（出处：常沙娜《敦煌历代服饰图案》，万里书店有限公司·轻工业出版社，1986年，图86）

峰造极的程度。莫高窟艺术中的卷草纹实例多见于边饰、圆光、服装，其作为唐代彩塑佛、菩萨、僧人服装的重要装饰纹样，明显受到世俗纹样的影响（图23）❶，具有浓郁的时代特征。

其三，彩条纹，又分为线形条纹与V形条纹，前者穿插于长短裙各种纹样之间，后者位于短裙花朵纹下方。彩条纹由不同色彩按照一定规律进行排列形成，富有秩序感和装饰性，在唐代广泛地用于纺织品的装饰。线形彩条纹可以对比吐鲁番阿斯塔那墓出土唐代彩绘绢画双童穿着的背带裤❷。V形彩条纹也见于第45窟彩塑迦叶穿着的僧祇支上边缘（图24），可见其应用于服装时多作长窄条表现。

本文通过分析莫高窟第45窟盛唐两尊彩塑菩萨像的服饰，得出以下三方面认识。第一，从服饰的搭配及形制整体考虑，其与中原北方地区石刻造像呈现一体化发展态势，时代统一性特征十分显著。第二，具体到胸饰璎珞、络腋等细部表现，其对当地初唐菩萨像服饰的承袭一目了然，体现出的地域性特征也不容忽视。第三，高发髻、服装色彩与纹样反映了唐代社会的审美观念和欣赏要求，生动诠释了佛教造型艺术的本土化进程。总之，莫高窟第45窟彩塑菩萨像的服饰凝聚了时代精华，融入了地方特色，成就了经典之作。

---

❶ 新疆维吾尔自治区博物馆藏。
❷ 周菁葆、孙大卫主编《西域美术全集2：绘画卷》，天津：天津人民美术出版社，2016年，第40页。

图23　吐鲁番阿斯塔那230号墓出土　唐代彩绘舞伎局部
（出处：周菁葆、孙大卫主编《西域美术全集2：
绘画卷》，天津：天津人民美术出版社，
2016年，第43页）

图24　敦煌莫高窟第45窟西龛北侧　盛唐彩塑迦叶
局部［出处：段文杰主编《敦煌石窟艺术·莫高窟
第四五窟附第四六窟（盛唐）》，南京：江苏美术
出版社，1993年，图13］

Analysis of the Costume on the Painted Sculptures Bodhisattva Dated to the High Tang Dynasty in Cave 45 at Dunhuang Mogao Grottoes

Qi Qingyuan

Dunhuang Mogao Grottoes have preserved more than 2,000 painted sculptures from the Sixteen States Period to the Western Xia Regime, which not only plays an important role in understanding the comprehensive grotto art, but also provides valuable data for the study of Chinese ancient clothing culture.

Cave 45 is the most representative cave of the high Tang Dynasty at Mogao Grottoes, and this paper selects two painted sculptures Bodhisattva as the research subject. The study mainly focuses on three aspects: hair bun and treasure crown, accessories and clothes, and making comparative analysis with stone carving Bodhisattva of the Tang Dynasty from the northern part of the Central Plains as much as possible while connecting the relevant examples of other caves at Mogao Grottoes. It is expected to sort out painted sculpture Bodhisattva clothing characteristics during the high Tang Dynasty at Mogao Grottoes from this example to general, and clarify the internal relationship with secular clothes.

1. Basic information

The high Tang Dynasty painted sculptures in Cave 45 of Mogao Grottoes are placed in the west niche. There are seven existing sculptures, which are configured according to symmetry principle. In the middle is Buddha in lotus position, on both sides are standing sculptures of disciples, Bodhisattvas and Maharāja-devas. If Vīra sculptures on both outsides the niche still remain, this combination of nine body sculptures consistent with Vairocana Buddha statue niche layout built in the second year of Shangyuan Period (675) in Fengxian Temple at Longmen Grottoes, Luoyang. The two Bodhisattva statues are relatively same in shape (Fig. 1 and Fig. 2), height 185cm, tied in a high hair bun, wearing chest keyūra, armlets and bracelets, with diagonal Luoye (络腋) around the upper body and long skirt and waist wrap on the lower body. Facing slightly down, the head, upper body and waist form a S-shaped design, looks graceful, gentle, healthy and full of life vitality. Bodhisattva of this style were popular during the Tang Wu Zhou period to the Kaiyuan period (690-741). In terms of the overall matching of clothing (Fig. 3), the Bodhisattva sculptures in Cave 45 of Mogao Grottoes are highly similar to many examples in Guanzhong area, except that they do not have long sashes, such as the stone sculptures in Qibaotai of Guangzhai temple in Chang'an dated to the Tang Wu Zhou period, and the cliff carvings in Yaowang mountain at Yao county dated to the Kaiyuan period (Fig. 4). If we consider the modeling details of clothing, the Bodhisattva sculptures in Cave 45 of Mogao Grottoes did not directly imitate Guanzhong area style, but was affected by various factors, they should be put into a broader perspective to examine.

Fig. 1 Painted sculpture Bodhisattva dated to the high Tang Dynasty in the south of the west niche in Cave 45 at Dunhuang Mogao Grottoes [source: Dunhuang Grotto Art · Cave 45 with Cave 46 (high Tang Dynasty), edited by Duan Wenjie, Nanjing: Jiangsu Art Publishing House, 1993, Fig. 31]

Fig. 2 Painted sculpture Bodhisattva dated to the high Tang Dynasty in the north of the west niche in Cave 45 at Dunhuang Mogao Grottoes [source: Dunhuang Grotto Art · Cave 45 with Cave 46 (high Tang Dynasty), edited by Duan Wenjie, Nanjing: Jiangsu Art Publishing House, 1993, Fig. 25]

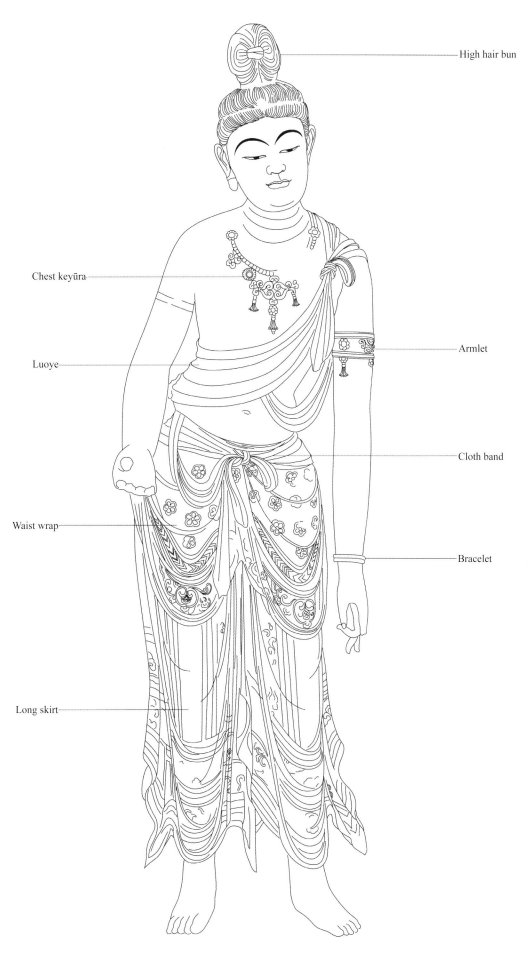

High hair bun

Chest keyūra

Luoye

Armlet

Cloth band

Waist wrap

Bracelet

Long skirt

Fig.3 The dress name of the painted sculpture Bodhisattva dated to the high Tang Dynasty on the north side of the west niche in Cave 45 at Dunhuang Mogao Grottoes (painted by Qi Qingyuan)

2. Hair bun and treasure crown

The two painted sculpture Bodhisattvas in Cave 45 of Mogao Grottoes both have a high hair bun and combed neatly. According to the hair shape, it can be divided into three parts: forehead hair, middle hair and top hair, as well as sideburns that droop to the ears (Fig. 5). This three-parts high hair bun is a typical feature of the Tang Dynasty Bodhisattva (Fig. 6). The high hair bun on the top of the head is tied in the middle with a strap and then fall to left and right sides into a folded ring. This was the most popular Bodhisattva hair bun style in the early and high Tang Dynasty, also common in grottoes, cliff carvings and unearthed sculptures from ground temples. The sideburns of the two Bodhisattvas are

Fig. 4 Double Avalokitesvara Bodhisattva statue at niche 9 of Yaowang mountain cliff dated to the 11th year of Kaiyuan of Tang Dynasty (723) (photo by Qi Qingyuan)

Fig. 5 The hair bun and treasure crown illustration of the painted sculpture Bodhisattva dated to the high Tang Dynasty on the south side of the west niche in Cave 45 at Dunhuang Mogao Grottoes (painted by Qi Qingyuan)

wrapped around the ears, and the sideburn of the better preserved south Bodhisattva around left ear is divided into two strands, these seem followed Avalokitesvara and Mahasthamaprapta Bodhisattva sculptures style dated to the Early Tang Dynasty on the west wall of the Big Buddha cave in the Dafosi Grottoes at Bin County (Fig. 7). This kind of design was not common in the Tang Dynasty, but gradually became popular in the Song Dynasty and beyond.

In the Tang Dynasty, women have high hair bun was a fashion. At that time, the popular Banfan bun (半翻髻, Half turned bun), Huihu bun (回鹘髻, Uighur bun), Jinghu bun (惊鹄髻, Startled Swan bun), Zhui bun (椎髻, Vertebral bun) and Luo bun (螺髻, Spiral bun) all belonged to high hair bun. Through the records of Tang poetry, we can further understand the popularity of high hair bun. For example, Bai Juyi's *Jiang Nan Xi Feng Xiao Jiuche Yin Hua Chang'an Jiuyou Xi Zeng Wu Shi Yun* (《江南喜逢萧九彻因话长安旧游戏赠五十韵》): "Now days high hair bun is a fashion, the makeup is light." Yuan Zhen's *Li Wa Xing* (《李娃行》): "The hair bun is a foot high, everybody is happy just like stand in spring wind." Although the specific form of Bodhisattva high hair bun in the Tang Dynasty is difficult to fully correspond to secular hair bun in real life, while this still reflect the aesthetic interest and the fashion at that time. It should be mentioned that the female musician in the Wei Guifei tomb dated to the second year of Tang Qianfeng (667) in Liquan, Shaanxi, the sideburns droop to ears, which provides some clues for the study of sideburns around ears for some Bodhisattvas in the early and the high Tang Dynasty.

The Bodhisattva sculpture on the south side of the west niche in Cave 45 at Mogao Grottoes wears a crown, which is decorated with two flower medallion jewels on the crown hoop, the upper edge decorated with a circle of beads, and a Cintamani placed on both left and right sides (same as Fig. 5). The combination of crown hoop and Cintamani appeared on painted sculptures and mural Bodhisattvas' crown at Mogao

Fig. 6 Tang Dynasty Bodhisattva's head in Cave 18 of Tianlongshan Grottoes in Taiyuan (photo by Qi Qingyuan)

Fig. 7 Avalokitesvara Bodhisattva of the early Tang Dynasty on the west wall of the big Buddha cave in the big Buddha Temple Grottoes in Bin county (photo by Qi Qingyuan)

Grottoes during the Sui Dynasty, which continued to develop in the early and high Tang Dynasty. However, most treasure crowns have a Cintamani in the middle, that forms a typical three jewels crown with ornaments on both sides. Considering the hair bun and treasure crown two factors, the similar shape and combination way can be compared with the Avalokitesvara Bodhisattva sculpture dated to the Tang Dynasty unearthed in Jinglongchi, Dongguan, Xi'an (Fig. 8), except there is a Buddha image in the middle of its treasure crown to show the identity. Like the Bodhisattva on the south side of the west niche in Cave 45 at Mogao Grottoes, this is very rare that only both sides of the crown have ornaments. Therefore, it can not be ruled out that the treasure crown was originally designed with Cintamani in the center or something indicates the identity. Unfortunately, it is not intact because long time passed. There is a circle of wear marks between the forehead and middle hair of the Bodhisattva on the north side of the west niche (Fig. 9), we believe the Bodhisattva should also has a crown originally.

Fig. 8 Part of the Avalokitesvara Bodhisattva statue dated to the Tang Dynasty unearthed in Jinglongchi, Dongguan, Xi'an (photo by Qi Qingyuan)

Fig. 9 Painted sculpture Bodhisattva head dated to the high Tang Dynasty in the north of the west niche in Cave 45 at Dunhuang Mogao Grottoes [source: Dunhuang Grotto Art · Cave 45 with Cave 46 (High Tang Dynasty), edited by Duan Wenjie, Nanjing: Jiangsu Art Publishing House, 1993, Fig. 26]

3. Accessory

The two painted sculpture Bodhisattvas in Cave 45 of Mogao Grottoes are similar to many stone sculptures in the northern part of the Central Plains, wearing chest keyūra, armlets and bracelets, which reflects the concise decoration characteristics of the Tang Dynasty Bodhisattva.

The chest keyūra of the two Bodhisattva sculptures are composed by neck ring and beads hanging connecting various ornaments (Fig. 10 and Fig. 11), which can be divided into main ornaments and secondary ornaments.

Fig. 10 Chest keyūra line drawing of the painted sculpture Bodhisattva dated to the high Tang Dynasty in the south of the west niche of Cave 45 at Dunhuang Mogao Grottoes (painted by Qi Qingyuan)

Fig. 11 Chest keyūra line drawing of the painted sculpture Bodhisattva dated to the high Tang Dynasty in the north of the west niche of Cave 45 at Dunhuang Mogao Grottoes (painted by Qi Qingyuan)

The main ornament of sculpture Bodhisattva chest keyūra on the south side is a huge flower medallion, which is composed by inner and outer parts. The inner part is beads around a large round jewel, and the outer part is six C-shape curves outwardly connected (there is a C-shape curve at the top, which only shows partially, seems to be limited by space). Their creativity and shape are similar to the flower medallion keyūra on the chest of Baoguan Buddha sculpture dated to the WuZhou period in the Leigutai Zhongdong of Longmen Grottoes in Luoyang. However, the latter six curves are connected in a inwardly direction (Fig. 12), and the form is slightly different. Many C-shape curves connected inwardly or outwardly into a circle, forming an important part of Baoxiang flower, a typical pattern of the Tang Dynasty (Fig. 13 and Fig. 14). In particular, usually there are six or eight C-shape curves as a group, from this we can see the source of C-shape curves flower medallion decoration of Bodhisattva images and treasure crown Buddha images of the Tang Dynasty.

The chest keyūra main ornament of the Bodhisattva sculpture on the north side is a pair of left-right symmetrical scrolling vine pattern. The scrolling vine pattern is in the shape of double stems with peach shaped head tend to center with thin neck, attached with inner curling small leaf buds. Similar scrolling vine ornaments are frequently seen in the whole body keyūra, chest keyūra and armlets of Bodhisattva sculptures in Cave 386, 205, 320, 319 and 328 (Fig. 15), this is a significant feature of painted sculpture Bodhisattva ornaments of the early and high Tang Dynasty at Mogao Grottoes. The creativity and shape of the scrolling vine ornament are similar to the double stems with peach shaped head tend to center with thin neck honeysuckle pattern of the Sui Dynasty and Northern and Southern Dynasties, which should be developed from this kind of honeysuckle pattern.

Fig. 12 Chest keyūra line drawing of the Baoguan Buddha sculpture dated to the Wu Zhou
period in the Leigutai Zhongdong of Longmen Grottoes in Luoyang (painted by Qi Qingyuan)

Fig. 13 The caisson center of Cave 334 at Dunhuang Mogao Grottoes
dated to the early Tang Dynasty [source: China Grottoes Dunhuang
Mogao Grottoes (Ⅲ), compiled by Dunhuang Institute of Cultural
Relics, Beijing: Cultural Relics Publishing House, 1987, Fig. 83]

Fig. 14 The caisson in Cave 320 dated to the high Tang Dynasty
at Dunhuang Mogao Grottoes [source: China Grottoes Dunhuang
Mogao Grottoes (Ⅳ), compiled by Dunhuang Institute of Cultural
Relics, Beijing: Cultural Relics Publishing House, 1987, Fig. 8]

The two Bodhisattva sculptures chest keyūra minor ornaments are small flower medallions and small beads around big jewels and tassels. They have been widely used in decoration on Bodhisattva since the late Northern Dynasty. In the early and high Tang Dynasty, chest keyūra with tassels was also common on stone statues of Longmen Grottoes in Luoyang and Tianlongshan Grottoes in Taiyuan, which can be compared with the Bodhisattva sculptures in Cave 45 of Mogao Grottoes.

The armlets of the two painted sculpture Bodhisattva are roughly same (Fig. 16). Between the two lines decorated with small flower medallions and C-shape curves outwardly connected into large flower medallions,

Fig. 15 Part of the painted sculpture Bodhisattva
dated to the high Tang Dynasty in the south of the
west niche of Cave 328 at Dunhuang Mogao Grottoes
(source: complete works of Chinese art · sculpture 7
Dunhuang painted sculptures compiled by Dunhuang
Research Academy, Shanghai: Shanghai People's Art
Publishing House, 1989, Fig. 126)

Fig. 16 Armlet line drawing of the painted sculpture
Bodhisattva dated to the high Tang Dynasty on the
north side of the west niche of Cave 45 at Dunhuang
Mogao Grottoes (painted by Qi Qingyuan)

with tassels hanging on the lower edge, indicating that the armlets and chest keyūra are based on the same design concept, and this feature was common on painted sculpture Bodhisattva ornaments during the high Tang Dynasty at Mogao Grottoes. The bracelets of the two Bodhisattvas are same (Fig. 17). They are in double ring form, which was a very popular bracelet style of the Tang Dynasty Bodhisattva. The simple design is

Fig. 17 Bracelet line drawing of the painted sculpture
Bodhisattva dated to the high Tang Dynasty on the
north side of the west niche of Cave 45 at Dunhuang
Mogao Grottoes (painted by Qi Qingyuan)

similar to the double ring silver armlet unearthed from the Tang Dynasty hoard in Xiaxinqiao, Changxing, Zhejiang Province.

It is necessary to mention that the chest keyūra and armlets of the painted sculpture Bodhisattva dated to the high Tang Dynasty on the north side of the central altar in Cave 205 of Mogao Grottoes (Fig. 18) combined the C-shape curves outwardly connected into large flower medallions with left-right symmetrical scrolling vines, as if they were reborn from the integration of the main ornaments of the two Bodhisattva sculptures in Cave 45.

4. Clothing

4.1 Clothing style

The two painted sculpture Bodhisattvas in Cave 45 at Mogao Grottoes are diagonal covered with Luoye on the upper body and wear long and short skirts on the lower body. This was the most popular clothing form

Fig. 18 Part of the painted sculpture Bodhisattva dated to the high Tang Dynasty on the north side of the central Buddhist altar in Cave 205 at Dunhuang Mogao Grottoes [source: China Grottoes Dunhuang Mogao Grottoes (Ⅲ), edited by Dunhuang Institute of cultural relics, Beijing: cultural relics press, 1987, Fig.123]

Fig. 19 Part of the Maitreya Bodhisattva statue dated to the 5th century unearthed at site Sarnath in Uttar Pradesh (photo by Qi Qingyuan)

of Bodhisattva in the Tang Dynasty, fashioned all over the country.

Luoye is a long piece of silk, which wrapped around the upper body and tied at the upper part of the left chest to form a design that covers from the left shoulder to the right abdomen. The two Bodhisattvas have similar Luoye but with difference, the shape of the ends after tying are different, which appears flexible and changeable. The longer part of the Luoye on the south side Bodhisattva sagging to the abdomen, and the end is stuffed into clothes in front of abdomen. The other part is wrapped behind the left shoulder (same as Fig.1). The longer part of the Luoye on north side Bodhisattva sagging from the left shoulder to behind, then winding to front, and the end is stuffed into clothes; the other part droops naturally after a single flower knot was tied on the left chest (same as Fig. 2 and Fig. 3). Looking at the Bodhisattva Luoye of the Tang Dynasty, many of them covered from the left shoulder to the right abdomen, and there are not a few covered from the right shoulder to the left abdomen. The former is obviously influenced by the Bodhisattva clothing of the Gupta Dynasty (Fig. 19), and the latter is likely formed in China. These two Luoye designs are both included in the Bodhisattva sculptures of the Tang Dynasty at Mogao Grottoes, which developed synchronously with the stone statues in the north of the Central Plains. However, in most cases, this two kinds of Luoye ends usually wrapped around the left shoulder and chest or the right shoulder and chest, and the practice of tying is sporadically seen in the stone statues of Tianlongshan Grottoes in Taiyuan, but they are very popular at Mogao Grottoes, such as Cave 322 and 332 dated to the early Tang Dynasty and Cave 66, 79, 264, 320, 444 and 458 dated to the high Tang Dynasty. Therefore, the Luoye knot can be regarded as the regional clothing feature of the painted sculpture Bodhisattva of the early and high Tang Dynasty at Mogao Grottoes.

The lower body of the two Bodhisattvas are covered by ankle length skirts and short aprons, and the inner and outer skirts are tied by strips. The strips are similar and interesting. The two parts of the south side Bodhisattva's strip sagging from the knot to form two U-shaped design, and the ends are stuffed into the skirt (same as Fig. 1). One part of the north side Bodhisattva strip forms a U-shaped end, which is stuffed into the skirt, and the other part is a single flower knot and naturally droops (same as Fig. 2 and Fig. 3). Two skirts with long inside and short outside was the typical style of the lower clothes for Tang Dynasty Bodhisattva sculpture. The practice of lace up outside skirt was popular in the northern part of the Central Plains from Wuzhou to Xuanzong around Kaiyuan period, mainly distributed in Guanzhong, around Luoyang, Taiyuan and Qingzhou. The two Bodhisattva's skirts and strips design in Cave 45 at Mogao Grottoes are particularly similar to the Tang WuZhou period stone sculptures unearthed from Xingyang Dahai Temple site (Fig. 20).

4.2 Clothing color and pattern

The clothing colors of the two painted sculptures Bodhisattva are mainly red and green, matched with black, blue and golden colors, forming a colorful and magnificent visual effect, showing the aesthetic taste of the high Tang Dynasty. The Luoye, long skirts grounding and lace are pomegranate like vermilion, calm and gorgeous. The short skirt grounding is close to emerald green, bright and pure. Vermilion and emerald green are contrasting colors, with high color saturation, strong contrast and dazzling. Red and green were the main colors for women's clothing during the Tang Dynasty. The combination and matching can be seen from time to time, which not only directly reflected in the textiles, paintings and pottery figurines unearthed from the Tang tombs (Fig. 21), but also described by a large number of Tang poetry works. For example, Wu Zetian's *Ruyi Niang* (如

Fig. 20 Guangxiang Bodhisattva dated to the Tang Dynasty unearthed at the site of Dahai temple in Xingyang (source: Henan Buddhist stone statues edited by Henan Museum and Wang Jingquan, Zhengzhou: Elephant Press, 2009, plate 22 of statues of the Tang Dynasty)

Fig. 21 Triple colored woman sitting figurines dated to the Tang Dynasty unearthed in wangjiafen village, Xi'an City (Osaka City Museum of art, "the Beauty of Tang Dynasty Women Exhibition", China Japan news agency, 2004, Fig. 39)

Fig. 22 Brocade pattern on painted sculpture Bodhisattva skirt dated to the Tang Dynasty in Cave 45 at Mogao Grottoes in Dunhuang (source: Chang Shana, Patterns of China Dunhuang Dresses and Adornments in Different Ages, Wanli Bookstore Co., Ltd. Light Industry Press, 1986, Fig. 86)

意娘) said, "If you don't believe that I shed tears recently because I miss you, open the box and have a look at the tears on my pomegranate skirt." *Wuhoushi Tongyao* (武后时童谣): "The fragrance of palace women in red and green dress drifts thousands miles away." Mao Wenxi's *Zanpuzi* (赞浦子) said: "The beauty wake up, lazily tie a pink band and walk slowly in an emerald dress." Yuan Zhen's *Wanyan Xiangting* (晚宴湘亭): "Dancers in long red skirts spin like flying, and singers in green long sleeved clothes sing beautiful songs." From this point of view, the clothing color of the Bodhisattva sculptures in Cave 45 of Mogao Grottoes conforms to the aesthetic trend of red and green in the Tang Dynasty, and decorated with patterns of black, blue, golden and other colors, they are particularly beautiful and gorgeous, may even better than the secular women's clothing.

The two painted sculptures Bodhisattva skirt patterns (Fig. 22) can be divided into three types.

First, five-petal flower pattern, which is fresh, simple and flexible, scattered on the short skirt. Flower pattern was a popular textile pattern during the Tang Dynasty, it has four petals, five petals and six petals. They are not only common patterns in secular clothes (same as Fig. 21), but also applied to monk's clothes. For example, the painted monk in the west niche of Cave 199 and the painted sculpture monk on the north side of the west niche in Cave 320 at Mogao Grottoes are all dressed in red and green five-petal flower pattern kasaya. In the early and high Tang Dynasty, many murals and painted sculptures of Bodhisattvas at Mogao Grottoes were decorated with a variety of small flower patterns, such as silk scarf, Luoye, apron and long skirt, reflecting people's sincere love for such patterns at that time.

Second, the scrolling vine pattern, with undulating slender wavy branches and leaves on the left and right, having curved petals at intersections, where decorated on the lower edge of short skirt and the edge of long skirt. Scrolling vine pattern ushered into a great development opportunity in the Tang Dynasty. Its various forms and wide applications reached the peak. Examples of scrolling vine pattern at Mogao Grottoes are mostly seen in edge decorations, round lights and clothes. As an important decorative pattern of painted sculpture Buddha, Bodhisattva and monk clothes in the Tang Dynasty, it was obviously influenced by secular patterns (Fig. 23), which has strong characteristics of the time.

Third, color stripes are divided into linear stripes and V-shape stripes. The former is interspersed with various patterns on long and short skirts, and the latter is located below the short skirts flower patterns. The color stripe pattern is formed by arrangement of different colors according to a certain law, which full of order and decorative. It was widely used in textiles decoration during the Tang Dynasty. The linear color stripes can be compared with the suspenders worn by two children in the colored silk painting dated to the Tang Dynasty unearthed from the Astana tomb in Turpan. V-shape color stripes can also be seen on the upper edge of

Sankaksika worn by Kasapa, a painted sculpture in Cave 45 (Fig. 24). We can see that when it's used on clothing, its often long and narrow.

Fig. 23 Part of the Tang Dynasty painted dancer unearthed from tomb 230 in Astana, Turpan (source: complete works of art in the Western Regions, Volume 2, edited by Zhou Jingbao and sun David, Tianjin: Tianjin People's Art Publishing House, 2016, P. 43)

Fig. 24 Part of painted sculpture Kasapa dated to the high Tang Dynasty on the north side of the west niche in Cave 45 at Dunhuang Mogao Grottoes [source: Dunhuang grotto art·Cave 45 with 46 of Mogao Grottoes (high Tang Dynasty), edited by Duan Wenjie, Nanjing: Jiangsu Art Publishing House, 1993, Fig. 13]

By analyzing the clothes of two painted sculptures Bodhisattva dated to the high Tang Dynasty in Cave 45 at Mogao Grottoes, this paper has three following understandings: First, considering the collocation and style of their clothes as a whole, they present an integrated development trend with stone sculptures in the north of the Central Plains, and the time unity characteristics is very significant. Second, when it comes to details such as chest keyūra, Luoye, their inheritance of the local Bodhisattva clothing in the early Tang Dynasty is clear, and the regional characteristics can not be ignored. Third, the high hair bun, clothing colors and patterns reflected the aesthetic concept and appreciation requirements for the Tang society, and vividly interpreted the localization process of Buddhist plastic arts. In short, the clothes of the painted sculptures Bodhisattva in Cave 45 of Mogao Grottoes condensed the essence of the time, integrated local characteristics and made these classic works.

Dunhuang Mogao Grottoes Cave 23 of the High Tang Dynasty

敦煌莫高窟盛唐 第23窟

　　第23窟是盛唐时期的代表洞窟之一，始建于盛唐时期的方形覆斗顶窟，经中唐、五代重修，清重修塑像。主室窟顶画莲花井心藻井，西披画弥勒经变，南披画观音普门品，北披画阿弥陀经变，东披画佛顶尊胜陀罗尼经变。西壁盝顶帐形龛内清塑一佛、二弟子、二菩萨。盝形龛顶中央画棋格团花，四披存十三身药师佛立像，五身供养菩萨。龛内西披和西壁清画土红底色，绘火焰纹佛光，南壁存一天王、三菩萨，北壁存中唐画一天王、二菩萨，南北壁的菩萨都被近代重描。龛沿两侧绘有精美的团花、海石榴卷草边饰。龛外两侧为白底，仅北侧上方存一地藏菩萨，下方存二菩萨。龛外南北各有一组台座，上面的塑像已失，北侧台座的东、南面存盛唐画男供养人三身。主室南壁、北壁、东壁画法华经变诸品。甬道顶存五代画降魔变，前室西壁门上存五代画垂幔、跌坐佛及一条西夏文划刻题记。

　　整窟的壁画题材以法华经变为主，壁画构思巧妙，按北壁、东壁、南壁的叙事顺序表现法华经变的情节，内容包括序品、药草喻品、譬喻品、常不轻菩萨品、见宝塔品、观音普门品、化城喻品。其中，药草喻品中画天空乌云密布，田间草木茂盛，农夫在田间耕作、驱牛犁地的场景，寓意"佛法如云，云降雨以润万物"。西壁龛内菩萨头戴宝冠，服饰精美，姿态优雅，手捧莲花。画师技艺高超，画面敷色和晕染层次丰富，人物造型准确，线条有力流畅，其中手部造型尤为生动，增添人物的真实感。壁画中的建筑写实细腻，结构清晰，反映当时的现实生活。

（文：杨婧嫱）

Cave 23 is one of the representative caves of the high Tang Dynasty. It was built in a square cave with a truncated pyramidal ceiling in the high Tang Dynasty, then repaired in the middle Tang Dynasty and the Five Dynasty, and the sculptures were repainted in the Qing Dynasty. The top of the main chamber is painted with a lotus in the center of the caisson, and Maitreya Sutra illustration on the the west slope, the Samantamukhaparivartah illustration on the south slope, Amitabha Sutra illustration on the north slope and Sarvadurgatipariśodhana Uṣṇīṣa Vijaya Dhāraṇī sūtra illustration on the east slope. Painted sculptures of one Buddha, two disciples and two Bodhisattvas are in the tent-shaped niche of the west wall which were repainted in the Qing Dynasty. In the niche's flat checks ceiling is filled with round flower pattern, and thirteen Bhaisajyaguru standing images and five offering Bodhisattvas images on the four niche's slopes. The west slope and the west wall in the niche are painted with earth red background and flame pattern Buddha light, one painted Maharāja-deva and three painted Bodhisattvas on the south wall, and one painted Maharāja-deva and two painted Bodhisattvas dated to the middle Tang Dynasty on the north wall. The Bodhisattvas on the north and south walls have been retraced in modern times. On both edges of the niche have exquisite round flower pattern, sea pomegranate scrolling vine pattern. The two sides outside the niche are white background, only one painted Ksitigarbharaja Bodhisattva is preserved above the north side and two painted Bodhisattvas are below. Outside the niche, there is one pedestal on each side, and the sculptures on them have been lost. There are three painted male donors dated to the high Tang Dynasty on the east and south side of the north pedestal. The south wall, north wall and east wall of the main chamber are painted with different chapters of Saddharmapundarika Sutra. There is Subdue Demons illustration dated to the Five Dynasty on the top of the corridor, and on the west wall above the door of the antechamber, there are paintings of valance and sitting Buddha which dated to the Five Dynasty, and a Western Xia inscription.

The content of the whole cave is Saddharmapundarika Sutra illustration as the subject. The mural arrangement is ingenious, which shows the Saddharmapundarika Sutra illustration by narrative sequence on the north wall, the east wall and the south wall. The content includes Ninaadaparivartah, Oshadhiparivartah, Opamyaparivartah, Sadaaparibhuutaparivartah, Stuupadarsanparivartah, Samantamukhaparivartah and Prvayogaparivartah. In Oshadhiparivartah illustration, dark clouds in the sky, various vegetation in the field, farmers farming and driving cattle to plough in the field implies that "Buddha's teachings just like rain, and rains on all things". The Bodhisattva in the west niche wears a treasure crown, who has exquisite clothes and elegant posture, and holds a lotus in His hand. The painter has superb skills that coloring and shading technique full of layers, accurate character modeling, powerful and smooth lines, especially the hand modeling, which added authenticity to the characters. The architecture in the murals is realistic and exquisite, with clear structure, reflecting the real life at that time.

(Written by: Yang Jingqiang)

The Bodhisattva's clothes
on the south side of the west niche in the main
chamber of Cave 23 dated to the high Tang
Dynasty at Mogao Grottoes

菩萨服饰
莫高窟盛唐第23窟主室西壁
龛内南侧

　　盛唐第23窟主室西壁龛内南侧持花菩萨，其形象眉目娟秀，广额丰颐，启唇欲语。身体姿态呈"S"形，丰腴雍容。左手捧举复瓣莲花，右手轻扶柳枝，纤细柔嫩的玉指，有触碰欲碎的感觉。

　　菩萨束髻头顶，余发披肩，佩戴宝冠，上半身披着络腋，络腋的正面为黄色，背面为绿色。下半身着红色长裙，裙子的两侧由璎珞固定兜揽其长度，以便于举步行走。腰间装饰有华美的彩绦。项链、臂钏、手镯应有尽有，各色彩带自肩垂下而随风飘动。这些充分体现出盛唐时期菩萨服饰艺术瑰丽多姿的美感。

（文：刘元风）

On the south side of the west niche in the main chamber of Cave 23 dated to the high Tang Dynasty, the Bodhisattva holding a flower who has beautiful appearance, wide forehead and round cheek, the mouth is slightly open seems wish to speak. The body posture is "S" shaped, plump and graceful. The left hand holds a double layered lotus and the right hand holds a willow branch, the fingers are slender and tender, just like made of jade and seems like they can be broken by a single touch.

The Bodhisattva has a hair bun on the top of His head and wears a treasure crown, the rest hair cover the shoulders. His upper body is covered with Luoye, the front side of the Luoye is yellow and the back side is green, and the lower body is wearing a long red skirt, and the two sides of the skirt are fixed by keyūra to hold, so it's more convenient for walking. The waist is decorated with colorful rope. Necklaces, armlets and bracelets are all set. The color bands hang down from the shoulders and flutter in wind. These fully reflect the magnificent beauty of Bodhisattva clothes art of the high Tang Dynasty.

(Written by: Liu Yuanfeng)

The farmer's clothes
(Farming in the Rain)
on the north wall of the main chamber
in Cave 23 dated to the high Tang
Dynasty at Mogao Grottoes

服饰 （雨中耕作图）农夫

莫高窟盛唐第23窟主室北壁

　　盛唐第23窟主室北壁的雨中耕作图作为法华经变药草喻品中的典型情景之一，犹如一幅反映当时社会生活气息浓郁的民俗画，画面中阴云密布，时雨普降，前面的农夫右手扶犁，左手扬鞭策牛在雨中耕作；后面的农夫肩挑麦捆正疾步冒雨赶路。两位农夫均头戴防雨席帽，耕作的农夫上身穿交领半臂衫，腰间系带，下着半长宽脚裤，衣服的领口、袖口和裤口部位均镶饰贴边，兼具耐用和装饰功能。挑担的农夫上穿"V"形领半臂布衫，下着长裤，腰间系有围裙。

　　画面的艺术处理上，人物造型生动简约，用线概括传神，色彩和谐自然，充分体现出盛唐时期乡间生活中耕作、收割的繁盛景象，及其通过画面所呈现出的诗情与画意。

（文：刘元风）

　　The Farming in the Rain on the north wall of the main chamber in Cave 23 dated to the high Tang Dynasty, as one of the typical scenes in the Oshadhiparivartah from Lotus Sutra illustration, is like a folk painting reflecting the strong flavor of social life at that time. In the picture, clouds are thick and the rain is falling. The farmer in front holds the plow with his right hand and whips the cattle to farm in the rain with his left hand. The farmer behind is running through the rain with a bundle of wheat on his shoulder. The two farmers wear rainproof hats, and the farming farmer wears a cross collar half-sleeved coat and ties at the waist; wide shorts on the lower body. The neckline, cuffs and pants cuffs are decorated with welts, which are both durable and decorative. The farmer carrying wheat wears a "V" shaped collar half-sleeved coat, trousers on the lower body and an apron at the waist.

　　In the artistic treatment of the picture, the characters are vivid and simple, lines are neat and lively, and the colors are harmonious and natural. This picture fully reflects the prosperous scene of farming and harvesting in rural life during the high Tang Dynasty, as well as the poetic and pictorial mood presented through the picture.

(Written by: Liu Yuanfeng)

敦煌莫高窟盛唐　第31窟

Dunhuang Mogao Grottoes Cave 31 of the High Tang Dynasty

第31窟始建于盛唐时期，经五代、清重修，为中小型覆斗顶窟。主室窟顶为团花卷草藻井，四周绘有垂幔。西披画法华经变中见宝塔品、从地涌出品，南披画普贤变，北披画文殊变，东披画法华经变中药草喻品、随喜功德品等。西壁开凿平顶敞口龛，清塑七身塑像。龛顶画说法图一铺，下画菩提华盖。龛内西壁画卷草火焰纹佛光，两侧各绘四弟子。龛沿两侧画半团花及龟背纹边饰。龛外南北侧各画一身立佛。南壁中间画卢舍那佛说法图，卢舍那佛胸部画须弥山，两侧环绕听法的菩萨、比丘、金刚、天王。北壁画报恩经变一铺，中部为说法场面，两侧按画面顺序展开故事情节。西侧画孝养品中须阇提太子割肉济父。东侧画恶友品中善友太子寻宝施财。东壁门上画释迦说法图一铺，门南、北以大壁面画立式帝释天各一铺。甬道中央画五代接引佛一铺。前室顶存部分五代说法图，西壁门上五代画七佛，门北五代画北方天王，门南五代画南方天王。前室南北壁画东部模糊，南壁画文殊变，北壁画普贤变。

此窟四披极具特点。西披出现新的元素，七宝塔前出现一对世俗男女供养人，从地涌出的菩萨位于水上平台上，说明此时法华经变开始表现释迦净土的元素。南北两披共同组成人物众多、构图复杂的普贤变及文殊变。东披画面中人物形象丰富，有许多反映盛唐时期的生活场景。东披下侧画母亲席地坐在屋子里，奶娘抱着婴儿在院子里逗乐的场景，还画有两位妇女相对，一位高举布偶，另一位伸手去取的玩耍场景。壁画中的妇女丰腴圆润，形似唐代画家周昉笔下的仕女。北披文殊变中，众供养菩萨、天女、天王、力士等虽然形象较小，但其服饰上大量使用各种小花图案进行装饰，体现出盛唐时期敦煌服饰图案达到繁花似锦的新阶段。

（文：杨婧嫱）

Cave 31 was built in the high Tang Dynasty, a medium size truncated pyramidal ceiling cave, and repaired in the Five Dynasty and the Qing Dynasty. The ceiling of the main chamber is painted round flower scrolling vine pattern caisson, surrounded by painted valance. The west slope is painted with the Stuupadarsanparivartah illustration and Budhisattvaprtiviivivarasamudgamaparivartah illustration of Saddharmapundarika Sutra. The south slope is painted with Samantabhadra Bodhisattva tableau, and Manjusri Bodhisattva tableau on the north slope, the Oshadhiparivartah and the Anumodanaapunyabhalanirdesaparivartah on the east slope, etc. A flat ceiling niche was carved in the west wall to have seven body painted sculptures made in the Qing Dynasty. The top of the niche is painted with a Dharma assembly, and the lower part is painted with a Bodhi canopy. In the niche, the west wall has scrolling vine flame pattern Buddha light, and four disciples are painted on both sides. The niche is decorated with half round flower pattern and turtle back pattern along the both sides. A standing Buddha is painted on the north and south sides outside the niche. In the middle of the south wall, Vairocana Buddha assembly is painted. Vairocana Buddha's chest is painted with Sumeru Mountain, surrounded by Bodhisattvas, Bhikkhus, Vajra Vīras and Maharāja-devas listening to Dharma on both sides. The north wall is painted with Ulambana Sutra of Mahavaipulya Buddha illustration, the middle is a Dharma assembly, and the story is unfolded on both sides according to the painting sequence. The west wall is painted the Chapter of Filial Piety, the Prince Sujati cut his own flesh to save his father. On the east wall is painted the Chapter of Bad Friends, which illustrated the good prince searching treasure to help people. Above the east door is painted Sakyamuni Dharma assembly, and the south and north walls each is painted with a standing Śakro devānām indrah respectively. In the center of the corridor, there is a painting of Reception Buddha dated to the Five Dynasty. On the top of the antechamber has Dharma assembly painted in the Five Dynasty. Seven Buddhas are painted above the west door, and the north side of the door is painted with Vaiśravaṇa, and Virūḍhaka is painted on the south side, which both dated to the Five Dynasty. The north and south walls in the antechamber are blurred partially, and the Manjusri Bodhisattva tableau on the south wall and the Samantabhadra Bodhisattva tableau on the north wall.

The four slopes in this cave are very special. New elements appeared in the west slope, in front of the seven-treasure pagoda has a male and a female donors. The Bodhisattvas coming out of the ground are located on the water platform, indicating that the Saddharmapundarika Sutra illustration began to appear the elements of Sakya Pure Land at this time. The north and south slopes together constitute the Samantabhadra Bodhisattva tableau and Manjusri Bodhisattva tableau with many characters and complex composition. Painting on east slope has lot of figures, and there are many scenes reflecting the life of the high Tang Dynasty. The lower part of the east slope has a mother sitting on the ground in the house, and the wet nurse holding and caring a baby in the yard. It also shows the playing scene of two women facing each other, one holding up a puppet and the other reaching for it. The women in the murals are plump and round, similar to the ladies depicted by Zhou Fang, a painter in the high Tang Dynasty. On the north slope is painted with Manjusri Bodhisattva tableau, although the images of Bodhisattvas, heavenly women, Maharāja-devas and Vajra Vīras are small, their clothes are decorated with a large number of small flower patterns, reflecting the new stage of colorful Dunhuang clothing patterns in the high Tang Dynasty.

(Written by: Yang Jingqiang)

The Bodhisattva's clothes
on the north ceiling slope of the main chamber
in Cave 31 dated to the high Tang Dynasty at
Mogao Grottoes

菩萨服饰
莫高窟盛唐第31窟主室窟顶
北披

盛唐第31窟主室窟顶北披有壁画一铺，壁画表现的是文殊菩萨与众多菩萨一起赴会的场景。其中，侍从菩萨的服饰有其特点，上半身内穿绿色络腋，外有天衣缠身，下半身着红色织锦长裙，下摆处有绿色贴边，另配有绿色内裙，内裙下摆处镶有红色贴边，使内外裙的色彩和谐有序，腰间垂以白色的打结裙带。

菩萨头戴珠宝镶嵌的箍头式花冠，项链、臂钏、手镯装缀其身。侍从菩萨其体态圆润，双手持以长柄香炉，跣足站于莲花之上。

在画面的艺术处理上，菩萨造型准确，面容端庄，手的指形柔细而优美。画面用线洒脱，疏密有致。色彩冷暖对比配置，层次分明。

（文：刘元风）

The north ceiling slope of the main chamber has a piece of painting in Cave 31 dated to the high Tang Dynasty. The painting shows the scene of Manjusri Bodhisattva attending the assembly with many Bodhisattvas. Among them, the clothes of the attendant Bodhisattvas have their own characteristics. Their upper body wears green Luoye and wrapped by silk scarf, and the lower body wears red brocade long skirt with green welt at the hem, and green inner skirt has red welt at the hem. The colors of the inner and outer skirts are harmonious and orderly, and knotted white belt hanging at the waist.

The Bodhisattva wears a headband inlaid with jewels, and His body is decorated with necklace, armlets and bracelets. This attendant Bodhisattva has round body shape, holding a long handled incense burner by both hands, and stands barefoot on the lotus.

In the artistic treatment of the picture, the Bodhisattva's shape is accurate, His face is dignified, and the fingers shape are soft, thin and beautiful. This picture's lines are free and smooth, density and sparsity are balanced. The color contrast is strong, clear in layers.

(Written by: Liu Yuanfeng)

The Bodhisattva's clothes
on the north side of the west niche top in the
main chamber of Cave 31 dated to the high
Tang Dynasty at Mogao Grottoes

菩萨服饰

莫高窟盛唐第31窟主室西壁

龛顶北侧

盛唐第31窟主室西壁龛顶说法图中的胁侍菩萨，双手捧复瓣莲花，端坐于莲花台上，双腿下垂，脚踩莲花。菩萨面庞丰满，高鼻秀目，俏丽而优雅，颈部有象征丰腴的唐代女性特有的三道折纹，头上佩戴花冠，花冠中央有高高翘起的藤蔓，犹如丹凤昂首，两侧有半开的莲苞，颈部和腕部佩戴串珠项链和手镯。

菩萨上身坦露，肩披天衣，天衣的正面为绿色，背面为红色，双色的彩带缠绕其身。下身穿赭褐色罗裙，罗裙下摆镶饰绿色的贴边。

在整体的艺术表达上，线条舒畅飘逸，色彩基调典雅，充分展现盛唐文化、艺术的开放性和多元化。

（文：刘元风）

On the top of the west niche in the main chamber of Cave 31 dated to the high Tang Dynasty, the attendant Bodhisattva in the picture sits on a lotus seat, with both hands holding a double layered lotus and stepping on a lotus cushion. This Bodhisattva has a plump face, high nose, beautiful eyes, who is beautiful and elegant. There are three folds on His neck which symbolizing woman's richness in the Tang Dynasty. He wears a corolla crown on His head, in the center of the crown, there are upturned vine decorations, just like a red phoenix holding its head high. There are half open lotus buds on both crown sides, and beads necklaces and bracelets around His neck and wrists.

The Bodhisattva's upper body is naked and His shoulders are covered with silk shawl, the front side is green and the back side is red. Two-color ribbons wrap around His body. The lower body wears an ochre brown gauze skirt, and the lower hem of the gauze skirt is decorated with green welt.

In the overall artistic expression, the lines are smooth and free, the color tone is graceful, which fully shows the openness and pluralistic styles of the high Tang Dynasty culture and art.

(Written by: Liu Yuanfeng)

图：刘元风　Painted by: Liu Yuanfeng

The Śakro devānām indrah's clothes on the north side of the east wall in the main chamber of Cave 31 dated to the high Tang Dynasty at Mogao Grottoes

帝释天服饰 莫高窟盛唐第31窟主室东壁 北侧

　　帝释天原为印度教神祇，后成为佛教护法神。犍陀罗艺术中的帝释天穿菩萨装或世俗装，手持金刚杵，有王者、武士的神格，这种武士造型在炳灵寺石窟、金塔寺石窟、莫高窟早期洞窟中都曾经出现。到了盛唐开凿莫高窟第31窟时，帝释天的形象已经与帝王更为近似了——体态丰腴，着曲领中单、上衣下裳、大带蔽膝，整体人物造型已经具有《历代帝王图》中帝王的特征。与此同时，这身帝释天的着装还融合了佛教天人元素——梳髻戴冠、璎珞遍身、披帛飞扬，在帝王的雍容华贵中融入了飘飘仙气，从而塑造了佛国世界的王者形象。

　　从目前壁画上仅存的五官线条判断，这身帝释天像是一位面容丰盈的女帝形象。由于在法海寺壁画等中国很多寺庙壁画中，帝释天都是女后或男人女相的少年帝王形象，因此在绘画整理时遵照当下在洞窟中所见更接近女装的造型，以女相表现。

（文：李迎军）

Śakro devānām indrah was originally a Hindu God, and later became a Buddhist Dharma protector. Śakro devānām indrah in Gandhara art wears Bodhisattva clothes or secular clothes, holds Vajra pestle, and has a divine personality of king and warrior. This warrior style has appeared in Bingling Temple grottoes, Jinta Temple grottoes and early caves of Mogao Grottoes. By the time, Cave 31 of Mogao Grottoes was excavated in the high Tang Dynasty, the image of Śakro devānām indrah is more similar to emperor — He is plump, wearing a curved collar Zhongdan, Yi on the upper body and Chang on the lower body, wide belt and Bixi in front of legs. The overall figure shape has the characteristics of the emperors in the painting of *Thirteen Emperors Scroll*. At the same time, the clothes of Śakro devānām indrah also has the elements of Buddhist celestial beings — hair bun and crown, keyūra and silk scarf, integrated the celestial feelings into the emperor's grace, then made a king's image of the Buddhist world.

Judging from the remaining facial lines on the mural, this body of Śakro devānām indrah looks like a plump female empress. In many Chinese temple murals such as Fahai Temple murals, Śakro devānām indrah is an empress or a young emperor with male body and female face. Therefore, when making this painting, the gender is expressed in female face according to the image closer to women in the cave.

(Written by: Li Yingjun)

The Vīra's clothes
on the north slope of the main chamber
in Cave 31 dated to the high Tang
Dynasty at Mogao Grottoes

力士服饰

莫高窟盛唐第31窟主室北披

力士是佛教的护法神，因手持金刚杵又名金刚力士。金刚杵源自古印度武器，因质地坚固、能击破万物而得名，进入佛教体系后成为力士的法器，以威猛的形态助力士降魔。

主室北披文殊菩萨赴法华会图里的这身面目狰狞的力士位于天王、孔雀明王下方，尽管头上的黑色颜料已经脱落，但依稀可以辨别其头发的形状。他头梳双髻，戴宝冠，赤裸上身，肌肉突起，身材健硕，左手紧握金刚杵，腰间缠裹锦裙，赤脚站于五彩祥云之上。力士身上戴的项圈、臂钏、手镯、璎珞等饰物丰富了服装的层次感，迎风飞扬的冠缯、披帛、腰带与怒目而视的神态、侧身挺立的姿势相得益彰，整体造型威猛雄健、极具张力。

（文：李迎军）

Vīra is the Dharma protector of Buddhism. He also known as Vajra Vīra because He holds a Vajra pestle. Vajra pestle originated from ancient Indian weapon, which is named for its hard material and ability to break through all things. After entering the Buddhist system, it becomes a magic weapon of Vajra Vīras and helps them subdue demons.

The ferocious Vīra in the painting of Manjusri Bodhisattva going to the Dharma assembly on the north slope of the main chamber is located below the Maharāja-deva and Mahamayuri. Although the black color on His head has fallen off, we can vaguely distinguish the shape of His hair. He has double hair buns, a crown, naked upper body, protruding muscles, strong figure, clenching a Vajra pestle in left hand, wrapped a brocade skirt around waist, and stands barefoot on the colorful auspicious clouds. The collar, armlets, bracelets, keyūra and other ornaments on the body could enrich the sense of layers of the clothes. The flying crown laces, silk scarf and belt complement with the angry look and standing upright posture, and the overall shape is powerful, vigorous and full of tension.

(Written by: Li Yingjun)

图：李迎军　Painted by: Li Yíngjun

The Vīra's clothes
on the south slope of the main chamber
in Cave 31 dated to the high Tang
Dynasty at Mogao Grottoes

力士服饰

莫高窟盛唐第31窟主室南披

在主室南披的赴会图中，普贤菩萨端坐于莲座上，前有菩萨、天女导引，后有天王、力士、夜叉等护卫，一众人物驾五色祥云步虚行空、共赴法华会。整铺壁画人物众多，疏密有序，构图紧凑，描绘传神。

位于队列后方的这身金刚力士怒目圆睁，肌肉暴起，左手五指张开，右手紧握金刚杵，愤然欲吼，气势可畏。力士的着装具有典型的盛唐时期力士造型特征：头梳双髻，戴宝冠，赤裸上身，腰间缠裹有边饰的锦裙，系飘逸的长腰带，身上戴项圈、臂钏、手镯，垂挂璎珞。这幅力士像中，在身前身后多层缠绕、垂曳的宽大披帛是造型的亮点，流动的曲线既中和了刚健的躯体造型，又增强了整体形态的动感，使得这身五彩祥云之上的金刚力士栩栩如生、呼之欲出。

（文：李迎军）

In the painting of attending the Fahua Dharma assembly on the south slope of the main chamber, the Samantabhadra Bodhisattva sits on the lotus seat, guided by Bodhisattvas and celestial women in front, and escorted by Maharāja-devas, Vajra Vīras and Yakṣas. This group of beings ride on five colored auspicious clouds and walk in the air to go to the Fahua Dharma assembly. The whole mural has many characters, well arranged, very lively.

At the back of the crowd, the Vajra Vīra's eyes widened, His muscles hefty, the five fingers of His left hand stretched, and His right hand clenches the Vajra pestle. He is angry and wants to roar, looks threatening. The clothes of this Vīra has the typical modeling characteristics of Vīras in the high Tang Dynasty: His head has double hair buns, wears crown, the upper body is naked, the waist is wrapped with a brocade skirt with trimmed hem, and fastened by long flying belt. The body is wearing a collar, armlets, bracelets, and hanging keyūra. In this painting, the broad draped silk scarf has multiple layers, and winding and dragging around the front and back of the body is the special part of the whole clothes. The smooth curves not only neutralizes the vigorous hard body shape, but also enhances the dynamic of the overall shape, making this Vajra Vīra on the colorful auspicious clouds lifelike and ready to come alive.

(Written by: Li Yingjun)

The Maharāja-deva's clothes on the north side of the east wall in the main chamber of Cave 31 dated to the high Tang Dynasty at Mogao Grottoes

天王服饰 莫高窟盛唐第31窟主室东壁 北侧

同帝释天一样，天王也原是印度教神祇，后被佛教吸收成为佛国世界护法神，壁画中的天王身着裲裆甲，胸甲与背甲在肩头用宽皮带、在胸前用束甲带、在腰间用皮带固定。裲裆甲形成于魏晋战乱时期，具有极强的防护功能，但在这身天王画像的甲衣上已经见不到甲片，取而代之的是饱满的图案与艳丽的色彩。此外，唐时期军队使用的裲裆甲是武装全身的一系列完整装备，这身天王仅着身甲、胫甲，显然不是为了追求甲胄的防护功能。作为护卫帝释天出行的天王，其着装的装饰性已经远远高于甲衣的实用性。

在这身天王造型中，裲裆甲之上还绘有一块自胸前至腹下的白色板状饰物，这条白板通过束甲带与腰带固定在服装上，这种造型与使用方式在莫高窟的天王及士兵铠甲形象中尚无近似情况发现。绘画整理时，按照目前在洞窟所见如实记录，其形制、功用有待研究。

（文：李迎军）

Like Śakro devānām indrah, the Maharāja-devas were originally Hindu Gods and later absorbed by Buddhism as a Dharma protector. The Maharāja-deva in the mural wears Liangdang armor. The chest armor and back armor are fixed by wide belts on the shoulders, fasten belts on the chest and a belt around the waist. Liangdang armor was formed during the war time of the Wei and Jin Dynasties and has strong protective function. However, there is no armor pieces on the Maharāja-deva portrait, instead, it is full of patterns and gorgeous colors. In addition, Liangdang armor used by the army in the high Tang Dynasty was a set of equipment for protecting the whole body. This Maharāja-deva only wears body armor and shin armor, obviously not to pursue the protective function. As Maharāja-deva who escorts Śakro devānām indrah's trip, the decoration function of His armor is much more than the protective function.

In the Maharāja-deva portrait, a white board ornament from the chest to the belly is painted on the Liangdang armor. The white board is fixed on the clothing by the armor fasten belt and waist belt. This style has not been found in other images of Maharāja-devas and soldiers' armor in Mogao Grottoes. When doing the illustration, it is copied according to what we can see in the cave at present, and its shape and function need to be studied.

(Written by: Li Yingjun)

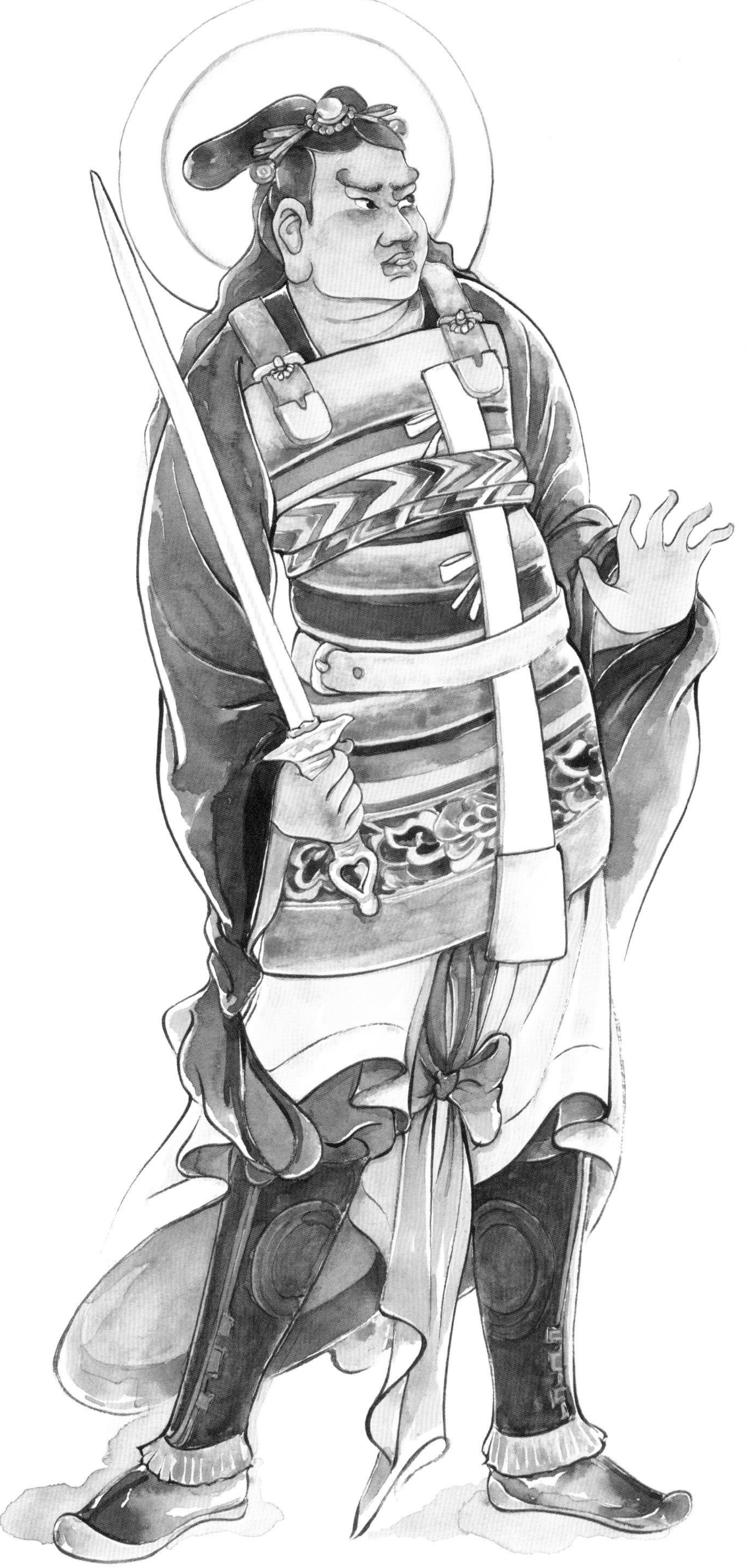

图：李迎军 Painted by: Li Yingjun

The offering Bodhisattva's clothes and pattern on the west side of the north slope in the main chamber of Cave 31 dated to the high Tang Dynasty at Mogao Grottoes

供养菩萨服饰及图案
莫高窟盛唐第31窟主室北披
西侧

　　盛唐第31窟主室北披绘文殊菩萨及其赴法华会中的供养菩萨，此图为位于文殊菩萨右前方的供养菩萨，手擎长茎青莲，后有天王、力士、夜叉以及孔雀明王护卫，脚踏祥云，步虚行空，共赴法华会。

　　菩萨青春面貌，梳低髻，戴宝冠。冠前中央镶宝珠，即摩尼宝珠。菩萨腕戴手镯、璎珞装身，璎珞由珠宝金玉等雕琢镶嵌、串联而成，故法华经变观音普门品有"解颈众宝璎珞，价值百万金而与之"之说。敦煌壁画、彩塑中菩萨及贵妇皆饰璎珞，早期样式简单，多作金玉环带，其上稍饰珠贝。到初、盛、中唐，则精雕细镂，金线编织，珠联玉串，绿松石镶嵌，网结连缀，长曳至腰，环垂腹前，极显高贵华丽之态。菩萨服饰色彩配置轻重相宜，上衣着半臂，搭饰披巾，垂落于身后，从图中的服饰造型来看，二者的材质都像是罗巾类的轻薄面料。菩萨腰身束蓝白两色的围腰，并在体侧系结，围腰之下穿装饰有四方连续散花纹的长裙。

（文：赵茜、吴波）

　　Manjusri Bodhisattva and the offering Bodhisattvas going to Fahua Dharma assembly are painted on the north slope of Cave 31 dated to the high Tang Dynasty. This painting is located in the right front of Manjusri Bodhisattva, holding a long stem green lotus in His hand, escorted by Maharāja-devas, Vajra Vīra, Yakṣa and Mahamayuri, standing on auspicious clouds and walking in the air.

　　This Bodhisattva has a youthful appearance, with a hair bun and a treasure crown. The central part of the crown is inlaid with a jewel, that is, Cintamani. The Bodhisattva wears bracelets and keyūra, and keyūra is made of processed gold, silver, jade etc., then inlaid and connected together into jewelry. Therefore, in the Saddharmapundarika Sutra-Samantamukhaparivartah, it is said that "The keyūra, which is a treasure untied from the neck of Bodhisattva, is worth millions of gold". In Dunhuang murals and painted sculptures, Bodhisattvas and noble women are decorated with keyūra. In the early stage, the style was simple, mostly gold and jade rings, with beads and shells on them. Then to the early, high and middle Tang Dynasty, they were carefully processed, woven together by gold thread, connected with beads and jades, inlaid with turquoise, and tied with nets, and elongated to waist and hung around the belly, showing a very noble and gorgeous feeling. The light color and dark color configuration of Bodhisattva's clothes is balanced. The upper body wears half-sleeved coat and draped silk scarf behind. From the clothing shape in the painting, the materials of both are light fabrics similar to Luo scarf. The Bodhisattva waist is wrapped with a blue and white waist wrap and tied at the side of the body. Under the waist wrap, He wears a long skirt decorated with four in a group repeated flower clusters pattern.

(Written by: Zhao Xi, Wu Bo)

图：吴波　Painted by: Wu Bo

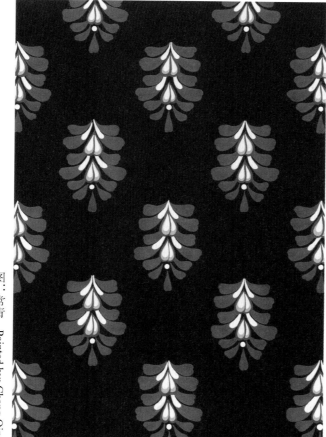

图：常青　Painted by: Chang Qing

The offering
Bodhisattva's clothes
on the left side of the north slope in the
main chamber of Cave 31 dated to the
high Tang Dynasty at Mogao Grottoes

供养菩萨
服饰
莫高窟盛唐第
31窟主室北
披左侧

此图为盛唐第31窟主室北披文殊菩萨赴法华会中的另一供养菩萨，也位于文殊菩萨右前方，与上述擎长茎莲花菩萨一起，前后导引，共赴法华会。

菩萨面部彩绘已失，轮廓呈青春貌。颈后部系有蝴蝶结，带缕颇长，垂于后背，根据菩萨衣饰推测，此"结"为菩萨天衣的一部分。菩萨双肩还搭饰仙帔，在敦煌艺术中，帔有两类，一类俗帔，即世俗人之披帛；另一类为仙帔（天帔），即菩萨、天王、力士之帔，这类帔随佛教艺术而来，受到中亚的影响。此菩萨梳低髻，头戴形制类似于当今发带的璎珞发饰，戴手镯、臂钏。菩萨下裳，腰束蓝、白、棕三色围腰，穿兼有蓝色小散花纹样的长裙，裙纹若隐若现间露出蓝色内裙。

（文：赵茜、吴波）

This painting shows another offering Bodhisattva who is in the group of Manjusri Bodhisattva going to Fahua Dharma assembly on the north slope of Cave 31 dated to the high Tang Dynasty. He is in the right front of Manjusri Bodhisattva, together with the above-mentioned Bodhisattva who holds a long stem lotus, they guide in front and back of the group to go to the Fahua Dharma assembly.

The Bodhisattva's face has faded, but the outline looks youthful. There is a bow tied at the back of the neck. The silk scarf is quite long and hangs on the back. According to the Bodhisattva's clothes, this "knot" should be part of the Bodhisattva's heavenly clothes. The Bodhisattva's shoulders are also decorated with Xianpei (仙帔). In Dunhuang art, there are two kinds of Pei, one is Supei, that is the common silk scarf for people; the other is the Xianpei (heavenly scarf), which is for Bodhisattvas, Maharāja-devas and Vajra Vīras. This kind of scarf came with Buddhist art and was influenced by Central Asia art. This Bodhisattva has a flat hair bun, a keyūra hair ornament with a shape similar to today's hair band, bracelets and armlets. The Bodhisattva wears Chang on the lower body, a blue, white and brown waist wrap at the waist, and a long skirt with small blue flower clusters pattern. The skirt is almost transparent, revealing the blue inner skirt.

(Written by: Zhao Xi, Wu Bo)

The secular mother and
daughter's clothes
on the north side of the east slope in
the main chamber of Cave 31 dated
to the high Tang Dynasty at Mogao
Grottoes

世俗母女
服饰
莫高窟盛唐第
31窟主室东
披北侧

此图位于盛唐第31窟主室东披北侧左下，表现的是法华经变七喻之一的"如子得母"，经文中颂扬伟大的母爱，众生得闻经义，就如儿女获得母爱一般。此图描绘了母亲与女儿之间戏玩木偶的温馨场面，再现了唐人母女的日常穿着和娱乐情景。

年长女性应为母亲，头梳垂鬟高髻，身穿窄袖襦裙。襦褐色，在襦之上，披帔，以线条勾勒结构但未着色，意在表现纱质面料的透明质感。其裙腰束得极高，正如唐周濆（濆）《逢邻女》诗："慢束罗裙半露胸"。母亲右手执木偶娃娃，左手微微提起长裙方便行走，露出褐色衬裙。女儿似豆蔻年华，梳两丸髻，脸上抹两圈脂粉（现已氧化变黑），同样穿着窄袖裙襦，襦石青色，裙为深色，一条轻薄透明的帔环绕于双臂之间，裙腰的束带与裙里均为石青色。

（文：董昳云、吴波）

This painting is located at the lower left of the north side of the east slope in the main chamber of Cave 31 dated to the high Tang Dynasty. It shows "just like a son gets his mother", which is one of the seven parables of the Saddharmapundarika Sutra. The scripture praises the great maternal love, and all sentient beings get the Scriptures, just like children get maternal love. This painting depicts the warm scene of playing puppet between mother and daughter, and reproduced the daily dress and entertainment of mother and daughter in the Tang Dynasty.

The older woman should be the mother, with her hair combed into Chuihuan high hair bun and wearing narrow-sleeved Ru skirt. The Ru is brown, and over the Ru is draped by silk scarf, and it has structure lines but not colored which is in order to show the transparent texture of the silk fabrics. The waistband of her skirt is tied very high, just like the poem written by Zhoufen in the Tang Dynasty *Meet Neighbor's Daughter*: "She wears Luo skirt slowly with breasts half exposed". The mother holds a puppet doll in her right hand and slightly lifted the long skirt in her left hand to help walking, revealing the brown petticoat. The daughter is very young, she has Liangwan hair bun and smeared two roundels of rouge on her face (now oxidized and blackened). She also wears a narrow sleeved Ru skirt, Ru is azurite, the skirt is dark color. A thin and transparent silk scarf is surrounded between her arms. The band at the waist and the inside of the skirt both are azurite.

(Written by: Dong Yiyun, Wu Bo)

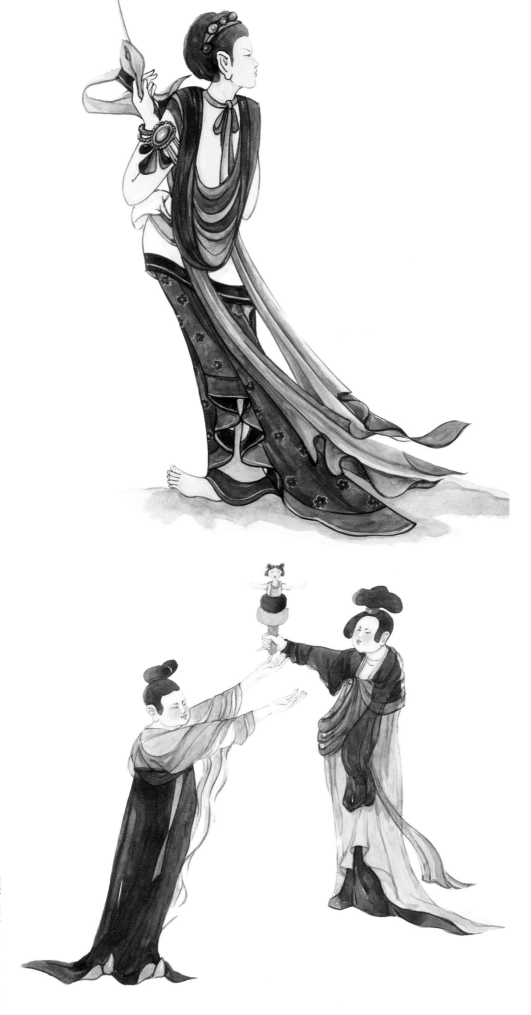

图：吴波　Painted by: Wu Bo

The celestial woman's clothes and dress pattern on the north slope in the main chamber of Cave 31 dated to the high Tang Dynasty at Mogao Grottoes

天女服饰及长裙图案

莫高窟盛唐第31窟主室北披

此图位于盛唐第31窟主室北披，为文殊赴会图中的一位供养天女，位于文殊菩萨和多位供养菩萨、天王之前，抬供桌，因此手并未露出。值得关注的是，该形象不似同壁其他供养菩萨那样衣着保留有明确的印度风貌——上身裸露，只带披帛，而是着交领襦，大袖，系腰带，下身长裙曳地，饰有唐代流行的小簇花，或为印花。这是典型的汉代以来的中国传统服饰。这样的服饰出现在壁画中，穿着者的身份等级相对较低，而同窟中大菩萨或佛的着装都是较传统的印度风格。换言之，大概也是主要的形象会相对传统、模式固定，可调整范围较窄，而其他随行人员似乎可以更加灵活地调整着装风格，适时地融入汉文化特征。长裙上的小簇花在莫高窟大范围出现是在中晚唐的供养人服饰上，而中原地区或唐都长安周边都是唐前期即有。

（文：张春佳）

This painting is located on the north slope of the main chamber in Cave 31 dated to the high Tang Dynasty. It is an offering celestial woman in the painting of Manjusri group attending the Dharma assembly. She is in front of Manjusri Bodhisattva and many offering Bodhisattvas and Maharāja-devas. She carries an offering table, so her hands are not showed. It is noteworthy that this image does not dress like other Bodhisattvas on the wall who have clear Indian style — the upper body is bare, only with silk scarf, but with a cross collar Ru, large sleeves, a belt, a long skirt trained on the ground, decorated with small clusters of flowers popular in the high Tang Dynasty, maybe printed. This is a typical traditional Chinese dress since the Han Dynasty. Such clothes appear in the murals, and the social status of the wearers are relatively low, while the clothes of Bodhisattvas and Buddhas in the same cave are more traditional Indian style. In other words, probably the main images are relatively traditional and fixed, and the adjustable range is narrow, while other accompanying people seem to be able to adjust their dress style more flexibly and could integrate the characteristics of Chinese culture in time. The small clusters of flowers on the long skirt appear in Mogao Grottoes on a large scale in the clothes of donors of the middle and late Tang Dynasty, while in the Central Plains or around Chang'an, the capital of the high Tang Dynasty, they appeared since the early Tang Dynasty.

(Written by: Zhang Chunjia)

图：张春佳　Painted by: Zhang Chunjia

图：常青　Painted by: Chang Qing

The Flying Apsaras' clothes
on the south wall in the main chamber of
Cave 31 dated to the high Tang Dynasty at
Mogao Grottoes

飞天服饰
莫高窟盛唐第31窟主室南壁

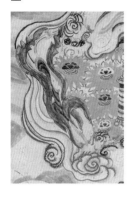

此图中的飞天位于盛唐第31窟主室南壁卢舍那佛上方，一组两身，围绕在华盖和圣树两侧，以云气烘托衔接整体造型。其中一身飞天，头戴璎珞，并饰有臂钏、项链、手镯，手持香炉，自上而下飞翔，腿部和衣裙向上飘举。飞天袒露上身，下着长裙。依据洞窟整体色彩状况，推断长裙应为红色调，裙摆处有绿松石色饰边。飘带与其他洞窟飞天相比更加宽大，画家在绘制时格外注意轻薄但有体量感的表现。第31窟建于盛唐末期，洞窟中的人物形态表现饱满，十分重视细节和服饰面料质感的呈现。但是飞天的璎珞和流云部分的绘画，已经与洞窟中其他部分的细节一同开始出现模式化的痕迹，接近部分中晚唐风格特征。同时，飞天的裙摆相比于第39窟等窟的飞天衣裙，已经略有简化，没有在末端加入更多的褶皱细节，整体上更为简洁。

（文：张春佳）

In this painting, Flying Apsaras are above the Vairocana Buddha on the south wall of the main chamber in Cave 31 dated to the high Tang Dynasty. A group of two bodies surround both sides of the canopy and the holy tree, supported by clouds. The left Flying Apsaras wears keyūra, armlets, necklace and bracelets, holding incense burner, flying up and down, and lifting up her legs and clothes. Her upper body is naked, and wears long skirt on the lower body. According to the overall color of the cave, we presume that the long skirt should be red, and the skirt hem has turquoise trim. Compared with other caves' Flying Apsaras, her ribbons are wider. The painter payed great attention to the performance of lightness out of volume. Cave 31 was built at the end of the high Tang Dynasty, the figures in the cave are comprehensive, and great attention was paid to the presentation of details and the texture of clothing materials. However, the paintings of Flying Apsaras' keyūra and clouds have begun to show stereotype together with the details of other parts of the cave, which are close to the style characteristics of middle and late Tang Dynasty partially. At the same time, the skirt of Flying Apsaras has been slightly simplified compared with the Flying Apsaras dress in Cave 39 and other caves, without adding more wrinkle details at the end, which is more simple on the whole.

(Written by: Zhang Chunjia)

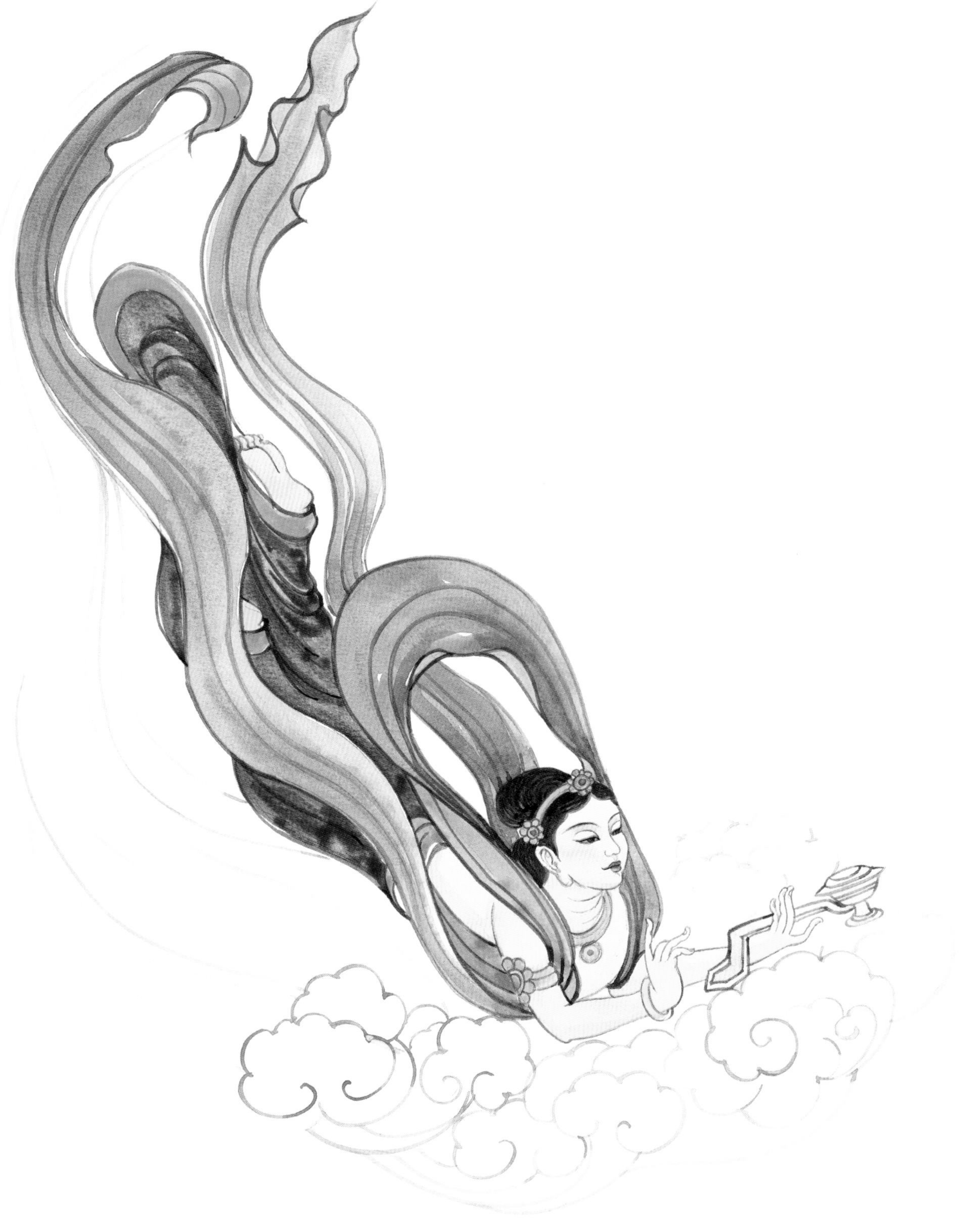

图：张春佳 Painted by: Zhang Chunjia

The Flying Apsaras' clothes
on the south wall in the main chamber of
Cave 31 dated to the high Tang Dynasty at
Mogao Grottoes

飞天服饰
莫高窟盛唐第31窟主室南壁

盛唐第31窟主室南壁卢舍那佛上方的右侧飞天，其裙腰部与左侧飞天有所区别，外翻部分为搭叠结构的石绿色，有饰边，目前看呈白色，但不排除为颜色褪色或剥落导致，长裙推测为红色调，裙摆有石绿色饰边，褶皱结构表现并不复杂，反而更多地将细节线条赋予了宽大的飘带，展现迎风飞舞、乘云而下的动人画面。飞天右手托盘，内盛鲜花，左手单手上举持花供养。以身姿和手持物件与左侧进行区分，力求在整体中呈现生动的变化。这两身飞天均无明显图案装饰，或有褪色剥落的可能，但从前面的供养天女和该窟其他部分的服饰可以看到，第31窟出现的服饰图案多为带叶小簇花，前面流行的六瓣、十字小团花此时已经不再大范围出现，这也是衔接后期洞窟服饰图案的一个体现。

（文：张春佳）

The waistband of the right Flying Apsaras who is located above Vairocana Buddha on the south wall of the main chamber in Cave 31 dated to the high Tang Dynasty is different from the one on the left. The upturned part is malachite green with overlapping structure and has trim. At present, it looks white, but it does not rule out that the color faded or fell off. The long skirt is speculated to be red, and the skirt has malachite green trim. The performance of the wrinkles is not complex, instead, the painter gave more detailed lines to the broad ribbons to show the vivid picture of riding on clouds and flying in wind. The Flying Apsaras' right hand holds a tray contains flowers, and the left hand holds a flower, different body posture and hand-held objects to distinguish from the left, and pursuing changes in the whole. There is no obvious pattern decoration for these two Flying Apsaras, or they may faded and fell off. However, we can see from the clothes of offering celestial woman mentioned before and other parts of the cave that most of the dress patterns in Cave 31 are small clusters of flowers with leaves, but the popular six-petal and small cross round flower pattern no longer appear on a large scale at this time, which is also a link connecting the dress patterns in later caves.

(Written by: Zhang Chunjia)

The wet nurse's clothes
on the north side of the east slope in the
main chamber of Cave 31 dated to the
high Tang Dynasty at Mogao Grottoes

乳母服饰

莫高窟盛唐第31
窟主室东披北侧

此图位于第31窟主室东披北侧，在母女戏玩木偶画面下方，表现了"如子得母"经变情节。母亲居中坐于一房屋内，奶娘抱着婴孩，在院子里边踱步边逗乐，展现出盛唐时期哺育婴孩的世俗场景与和睦慈爱的家庭氛围。

奶娘头梳抛家髻，身穿浅色窄袖上襦，下着深色长裙，肩头似垂下一条朱砂色披帛。怀中婴孩身穿朱丹色"裲裆"形制背心。画师在描绘奶娘的体态时，将其身体微微后仰，突出前腹，给怀抱婴儿适合的着力重心，她一只手抚住婴孩后背，似在轻轻拍打，另一只手稳稳托住婴儿的身体，表现出娴熟的育儿姿态。

（文：董昳云、吴波）

This painting is located on the north side of the east slope in the main chamber of Cave 31, below the painting of mother and daughter playing puppet, it also shows the classic plot of "like a son gets his mother". The mother sits in the middle of a house, and the wet nurse holds the baby, pacing and amusing in the yard, showing the secular scene of feeding babies and the harmonious and loving family atmosphere in the high Tang Dynasty.

The wet nurse combs a Paojia hair bun, wears light colored narrow-sleeved Ru on the upper body and a dark long skirt on the lower body, a cinnabar silk scarf is draped over her shoulders. The baby in her arms is wearing a red "Liangdang" shaped vest. When depicting the body of the wet nurse, the painter tilted her body slightly back and protruded her front belly to give her a suitable posture for holding the baby. She holds the baby's back with one hand, as if she is petting gently, and the other hand support the baby's body firmly, showing a skilled parenting posture.

(Written by: Dong Yiyun, Wu Bo)

The man's clothes
on the north slope of the main
chamber in Cave 31 dated to the high
Tang Dynasty at Mogao Grottoes

男子服饰

莫高窟盛唐第31
窟主室北披

莫高窟第31窟主室北披的右下方，四身士庶阶层的男装形象颇具特色。四人皆穿着圆领襕衫，头戴幞头，腰束革带，足蹬乌皮靴。其中，最左侧跪姿男子穿白色襕衫，戴折脚幞头，另三人均为站姿，戴长脚幞头，三人之中两人着绿色襕衫，一人着红色襕衫，襕衫的衣身宽松肥大，衣长及地。

幞头在唐初开始流行全国，当时的幞头多以轻薄柔软的纱罗制成，系结并垂在脑后的两个脚也是柔软悬垂的，被称为软脚幞头。初唐以后，垂于脑后的两脚开始逐渐加长，于是流行长脚幞头。莫高窟第31窟法华经变画中的男子所戴的幞头两脚长如带，直垂至肩前侧，正是典型的长脚幞头造型，这一形象也是当时流行服饰的真实反映。

（文：李迎军）

At the lower right of the north slope of the main chamber in Cave 31 at Mogao Grottoes, there are four men's clothes quite interesting. They all wear round collar Lanshan, Fu hat, leather belt and black leather boots. Among them, the man in the kneeling position on the left is wearing white Lanshan and bend-ends Fu hat. The other three are standing posture, wear long-ends Fu hat. Two of the three are wearing green Lanshan and one is red, the Lanshan are loose and wide, floor-length.

Fu hat became popular all over the country in the early Tang Dynasty. At that time, Fu hat were mostly made of thin and soft gauze, and the two ends tied and hanging behind head were soft, which was called soft ends Fu hat. After the early Tang Dynasty, the ends hanging at back of head began to elongated gradually, so the long ends Fu hat was popular. The Fu hat worn by the man in the Saddharmapundarika Sutra illustration painting in Cave 31 of Mogao Grottoes which the ends as long as a band and hanging straight to the front of the shoulders. It was a typical long ends Fu hat style, and this image is also a true reflection of clothes fashion at that time.

(Written by: Li Yingjun)

图：吴波　Painted by: Wu Bo

图：李迎军　Painted by: Li Yingjun

The donors' clothes
on the west slope of the main chamber
in Cave 31 dated to the high Tang
Dynasty at Mogao Grottoes

供养人服饰

莫高窟盛唐第31窟主室

西披

此图位于第31窟主室西披，主要表现《法华经》中见宝塔品虚空会经变故事，与东披的"灵鹫会"遥相呼应。绘释迦、多宝二佛并坐七宝塔内，宝塔"从地涌出，住虚空中"，佛家称为"虚空会"。七宝塔前有一对世俗男女供养人，相对而跪，中间题榜漫漶，两人双膝下绘祥云，作揖行礼。

男供养人双手合十，头戴幞头，两脚稍长，身穿褐色圆领缺胯袍。袍从胯部以下开衩，最早来自军衣，方便士兵骑马打仗，后因开衩的下摆非常适合骑马和劳动，为百姓所喜爱，遂成为庶民百姓的日常穿着。据《文献通考》载中书令马周上议："《礼》无服衫之文，三代之制有深衣。请加襕、袖、褾、襈为士人上服。开骻者名曰缺骻衫，庶人服之。"女供养人双手执团扇作揖礼状，头梳高髻，戴耳环，面颊有两圈胭脂晕，身穿对襟大袖襦裙，袖口宽广，有一圈浅色袖祛，襦长至臀下。男女二人神态谦卑，虔诚礼佛，一心供养。

（文：董眹云、吴波）

This painting is on the west slope of Cave 31. It mainly shows the story of "Dharma assembly in void" in Stuupadarsanparivartah of the Saddharmapundarika Sutra, echoing with the "vulture peak Dharma assembly" on the east slope. Buddha Sakyamuni and Prabhutaratna Buddha sit side by side in the seven-treasure pagoda. The pagoda "appeared from ground and stays in the air", which is called "Dharma assembly in void" by Buddhists. In front of the seven-treasure pagoda, there are a pair of secular male and female donors, kneeling to each other on auspicious clouds and bow to each other. The inscription in middle already faded.

The male donor puts his hands together, wears brown round collar slit robe, a Fu hat with slightly longer ends. The robe has two slits from waist to hem, which was first introduced from military uniform, for soldiers easier to ride and fight. Later, because the slit hem was very suitable for riding and working, so they are favored by ordinary people, and became the daily wear of the common people. According to Ma Zhou's suggestion in the book *Comprehensive Textual Research of Historical Documents*: "The book of *Rites* has no records of clothes, and in the previous generations have long clothes. Please add Lan, Xiu, Biao, Zhuan as the noble's upper clothes. The clothes which open the two lower sides are called as Quekuashan, which are worn by the common people." The female donor holds a round fan by both hands to make a bow, combs her hair in a high bun, wears earrings, has two roundels of rouge on her cheeks, wearing a pair of large-sleeved Ru skirt with wide cuffs which has a circle of light colored Xiuqu, and the Ru skirt is long below her hips. Both the man and the woman look humble and devout.

(Written by: Dong Yiyun, Wu Bo)

图：吴波　Painted by: Wu Bo

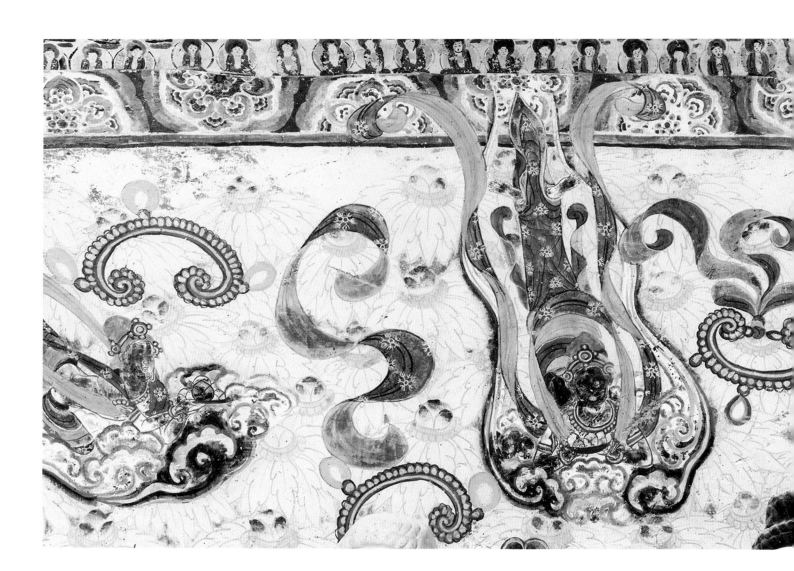

敦煌莫高窟盛唐

第39窟

Dunhuang Mogao Grottoes Cave
39 of the High Tang
Dynasty

　　第39窟是建于盛唐时期的人字披中心塔柱窟，经五代、西夏、清重修。前部人字披顶及后部平顶画千佛。主室中心柱东向面开斜顶敞口龛。龛内塑一佛及二弟子（清修），清塑二弟子。龛顶画说法图，周围有祥云环绕。龛壁画火焰纹团花佛光，四周画八大弟子。龛沿画菱格纹，龛外两侧画天王各一身，龛下五代画二立佛、四菩萨。中心柱南向面、西向面、北向面皆画千佛，南向面底部五代画壶门伎乐三身。主室西壁开凿斜顶敞口龛，彩塑涅槃像一铺二十七身，经清代重修。西壁龛顶及龛壁画五身飞天及四组天女。龛沿画半团花边饰，龛外两侧画千佛，龛下画垂幔，存五代画供养人七身。南北壁前部开斜顶圆券龛，龛外及后壁画千佛。南壁龛内塑一佛（清修）及清塑二佛。龛顶画菩提华盖，两侧各一身散花飞天。北壁龛内清塑三身，龛下存五代男供养人十八身。东壁门南北分别为五代画文殊变及普贤变，下画供养比丘，门上画千佛。前室存西夏画菩萨及净土变。甬道盝形顶五代画佛教史迹画，两披五代画瑞像图，南北壁西夏画供养菩萨。

　　此窟西壁龛顶及龛壁共画五身献花飞天，画四组梵天女奉摩耶夫人奔丧，均衡分布在龛顶及龛壁。飞天及天女形象受到唐代宫廷舞蹈和仕女画的影响，表现出世俗化的特征。飞天手持花蕾，各有彩云相托，随风而降，层层飘带随风长曳，延伸出灵动飘逸的气韵。天女姿态优美、雍容华贵、形象鲜明，犹如盛唐时期妇女的写照。整窟人物造型生动、线条细腻，体现出唐代工笔仕女画的风格。

（文：杨婧嫱）

Cave 39 is a central pillar cave with gabbled ceiling built in the high Tang Dynasty, which was repaired in the Five Dynasty, Western Xia and Qing Dynasty. Thousand Buddhas are painted on the front gabbled ceiling and the rear flat ceiling. The east side of the central pillar of the main chamber has a niche with an inclined top. One Buddha and two disciples painted sculptures are in the niche (repaired in the Qing Dynasty), and two painted disciples made in the Qing Dynasty. The top of the niche is painted with a Dharma assembly, surrounded by auspicious clouds. The murals of the niche are flame pattern, round flower pattern Buddha light, and eight disciples around. The edge of the niche is painted with rhomboid pattern, and the two Maharāja-devas are painted on both sides outside the niche. Below the niche is painted with two standing Buddhas and four Bodhisattvas dated to the Five Dynasty. The central pillar is painted with Thousand Buddhas on the south, west and north faces, and the bottom of the south side is painted Kunmen with three bodies of musicians dated to the Five Dynasty. The west wall of the main chamber has a niche with inclined ceiling, it is Nirvana theme with 27 bodies of painted sculptures, which were repaired in the Qing Dynasty. The ceiling of the west niche and the niche walls have five bodies Flying Apsaras and four groups of celestial women. Semi-round flower pattern is painted along the niche, Thousand Buddhas are painted on both sides outside the niche, valance is painted below the niche, and the seven donor images dated to the Five Dynasty. The outer part of the north and south walls have inclined top round edge niche, and the murals outside the niches and inner walls are painted with Thousand Buddhas. There is a painted sculpture Buddha in the south niche (repaired in the Qing Dynasty) and two Buddhas made in the Qing Dynasty. The top of the niche is painted with a Bodhi canopy, with Flying Apsaras scattering flowers on both sides. In the north wall niche, there are three bodies of painted sculptures made in the Qing Dynasty, and below the niche there are eighteen male donors dated to the Five Dynasty. The north and south of the east gate are the Five Dynasty painting of Manjusri Bodhisattva tableau and Samantabhadra tableau respectively, the lower is painted with offering Bhikkhu, and above the door is painted with Thousand Buddhas. In the antechamber, there are painted Bodhisattvas and Pure Land illustration dated to the Western Xia period. The flat corridor ceiling is painted with Buddhist historical stories dated to the Five Dynasty, and two slopes are painted with Auspicious Images dated to the Five Dynasty. On the north and south walls are painted with offering Bodhisattvas dated to the Western Xia period.

On the west niche ceiling and niche walls of this cave, five bodies flower offering Flying Apsaras are painted, and four groups of Brahma women are painted to serve Mahamaya to mourn, which are evenly distributed on the niche ceiling and niche walls. The images of Flying Apsaras and celestial women were influenced by court dance and lady painting in the high Tang Dynasty, showing the characteristics of secularization. The Flying Apsaras hold flower buds, ride on colorful clouds, descending with wind, and layers of ribbons flying, which have a free and elegant charm. The celestial women are graceful, elegant and vivid, just like portrait of women in the high Tang Dynasty. The figures in the whole cave are vivid and delicate, reflecting the traditional Chinese realistic painting style, lady paintings in the high Tang Dynasty.

(Written by: Yang Jingqiang)

The celestial woman's
clothes
on the west niche in the main chamber
of Cave 39 dated to the high Tang
Dynasty at Mogao Grottoes

天女服饰
莫高窟盛唐第39窟主室西壁
龛内

此图位于第39窟主室西壁龛内，表现佛母摩耶夫人惊悉佛陀涅槃，与三位天女急忙从忉利天宫前往娑罗双树间的情景。众人乘云而行，摩耶夫人焦急地注视着佛涅槃之处，身旁的天女回头凝望摩耶夫人，好似在劝慰佛母。

众人均梳惊鹄髻，装饰宝珠，发髻状如飞鸟的羽翼，作凌空欲飞之势。额前贴花钿，花钿为一种额饰，形状多种多样，有桃花、梅花、宝相花、菱形、圆形等形状，颜色为红、黄、绿等。另外，在众天女的面颊靠近太阳穴的位置，描有斜红，形如落叶，有时还故意描绘成残破状，宛如伤疤一样。在唐代，斜红是一种十分流行的妆饰。佛母摩耶夫人身穿半袖襦裙，饰有云肩，形制如唐代吴道子《天王送子图》中王后的云肩式样。云肩石绿色，边缘有曲线变化，形似卷云，两端上翘。佛母、天女的妆容服饰，应是以大唐宫妃写照为本。

（文：董昳云、吴波）

This painting is located in the west niche of Cave 39. It shows the scene that Mahamaya, the mother of Buddha was shocked to learn about the Buddha's nirvana and hurried with three heavenly maids from Tavatinsa to the Sal trees. When they move on clouds, Mahamaya anxiously looked at the nirvana site of Buddha. The celestial woman next to her turned back and stared at Mahamaya, as if comforting the Mother of Buddha.

All of them are combed in Startled Swan hair bun and decorated with jewels. The hair buns look like wings of bird, just like ready to fly in the air. Their foreheads are applied Huadian, which is a kind of forehead ornament with various shapes, including peach blossom, plum blossom, Baoxiang pattern, diamond, circle and other shapes, and the colors are red, yellow, green and so on. In addition, the cheeks near the temples of the celestial women are painted with Oblique Red, shaped like fallen leaves, and sometimes deliberately draw in incomplete shapes, like scars. In the high Tang Dynasty, Oblique Red was a very popular make-up. Mahamaya wears a half-sleeved Ru skirt decorated with cloud shoulders, and the shape is like the cloud shoulders of the queen in the painting of *Birth of Gautama Buddha* by Wu Daozi dated to the high Tang Dynasty. The cloud shoulders are malachite green, with curve changes at the edge, like cirrus clouds, and both ends are upturned. The make-up and clothes of the Buddha Mother and the celestial woman should be based on the imperial concubine portrait of the high Tang Dynasty.

(Written by: Dong Yiyun, Wu Bo)

The Flying Apsaras'
clothes
on the west niche in the main chamber of
Cave 39 dated to the high Tang Dynasty at
Mogao Grottoes

飞天服饰
莫高窟盛唐第39窟主室西壁
龛内

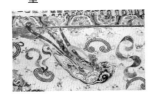

盛唐第39窟的飞天群组位于主室西壁龛内，该窟形制为中心柱窟，西壁佛龛位于中心柱后部，龛内塑涅槃佛像。龛顶共绘5身飞天，均持花向佛像飞来。上图飞天位于龛顶中部左侧，飞翔方向自左上向右下。飞天乘流云而下，飘带向后飞扬，并且十分生动的是，画家将裙幅和飘带的末端绘制在西龛内侧边饰上，与半团花形成压叠之态，生动有趣。飞天上身袒露，以飘带环绕，佩戴头饰、颈饰、臂钏、手镯等饰物，托举莲花，下着长裙，腰部翻出并以带系扎，裙身和飘带分别饰有六瓣和十字结构的小团花。裙摆处露出两足，褶皱表现较为复杂细致。对于飘带和裙腰、裙身的表现都讲求层次感。这样的层次表现使画面呈现出一定的空间感。说明相对于前朝洞窟，莫高窟唐代壁画尤其是盛唐时期，较为追求写实和空间层次感，洞窟整体的艺术倾向带动其中的局部人物表现。服饰色彩较为饱满，目前所见的壁面颜色以土红、石绿为主，对于氧化变色部分的色彩依据周边进行推测，呈现出较为华丽的气氛。

（文：张春佳）

The Flying Apsaras group in Cave 39 dated to the high Tang Dynasty is located in the west niche of the main chamber. The cave is shaped as a central pillar cave. The west niche is located at the back of the central pillar, and the nirvana Buddha sculpture is in the niche. Five Flying Apsaras are painted on the top of the niche, all flying toward Buddha with flowers. The Flying Apsaras in the picture is located on the middle left of the niche top, and the flying direction is from top left to bottom right on clouds, the ribbons flying in the back. What is very special is that the painter painted the skirt and the end of the ribbons on the inner side of the west niche and overlapped with semi-round flower pattern, which is vivid and interesting. The Flying Apsaras' upper body is naked and surrounded by ribbons. She wears head wear, necklace, armlets, bracelets and other ornaments, holds a lotus, wears a long skirt, which turned out at the waist and tied with a belt. The skirt and ribbons are decorated with small flower clusters with six-petaled and cross-structured respectively. The feet are exposed at the hem, and the folds are complex and detailed. To depict ribbons, waist wrap and skirt, we payed attention to the sense of layers. Such layer expression makes the painting showing a certain sense of space. Compared with the caves of previous dynasty, the murals of the Tang Dynasty in Mogao Grottoes, especially in the high Tang Dynasty, pursued realism and spatial layers, and the overall artistic tendency of the caves leads the performance of characters. The clothing color is relatively saturated, at present, the wall color is mainly reddish-brown and malachite green. The color of the oxidation and discoloration parts are speculated according to the surroundings, showing a gorgeous atmosphere.

(Written by: Zhang Chunjia)

右侧飞天飞翔的姿势与龛顶左侧飞天相对，大体对称。飞天头部、颈部、手臂、手腕等部分均佩戴首饰，双手托举盛有莲花蕾的花盘。裙腰部分翻出的层次以土红色和石绿色搭配，腰带部分系单结并垂下飘带。裙摆部分的绘制注重表现飘摆的褶皱和翻卷的动态，并且整体感觉较为窄瘦。考虑到实物和人体活动的状态，并且参照印度同时期的佛教人物服饰，这种造型应是画家夸张、理想的手法所致，与实物会有一定的出入。穿插于手臂部分的长飘带形态同前，正面为土红底色上印六瓣小团花，飘带反面也为石绿色。飞天乘云而下，云气纹也成为重要的飞翔辅助表现元素，流动方向与人物形态和飘带的方向协调一致。总观这三身飞天，飞翔在涅槃佛龛顶，并且飘带和裙幅飞出本部分空间，搭叠于半团花龛边饰上，这样生动的手法展现了盛唐画家的创新精神和活跃的艺术氛围。

（文：张春佳）

The Flying Apsaras flying posture on the right side is opposite to that on the left side of the niche top, which is roughly symmetrical. The Flying Apsaras wears jewelry on the head, neck, arms and wrists, and holds the flower plate containing lotus buds by both hands. The layers turned out at the waistband of the skirt are matched with reddish-brown and malachite green, and the belt is tied with a single knot and hanging with ribbons. When drawing the skirt the painter payed attention to the dynamic performance of the flowing folds and rolls, and the overall feeling is narrow and thin. Considering the real clothes and human bodies, and referring to the Buddhist figure clothes of the same period in India, this modeling should be the artist's exaggerated ideal, which will be different from the reality. The long ribbon interspersed with arms is the same as before, the front side is reddish-brown, with six-petaled small round flowers pattern, and the back side of the ribbon is also malachite green. The Flying Apsaras rides on cloud, the cloud pattern has also become an important auxiliary expression element of flying, and the flow direction is consistent with the character shape and the direction of the ribbon. In general, these three bodies Flying Apsaras, who are flying on the top of Nirvana niche, and ribbons and skirts reached out and overlapped on the semi-round flower pattern edge decoration of the niche. Such vivid technique shows the creative spirit and active artistic atmosphere of the high Tang Dynasty painters.

(Written by: Zhang Chunjia)

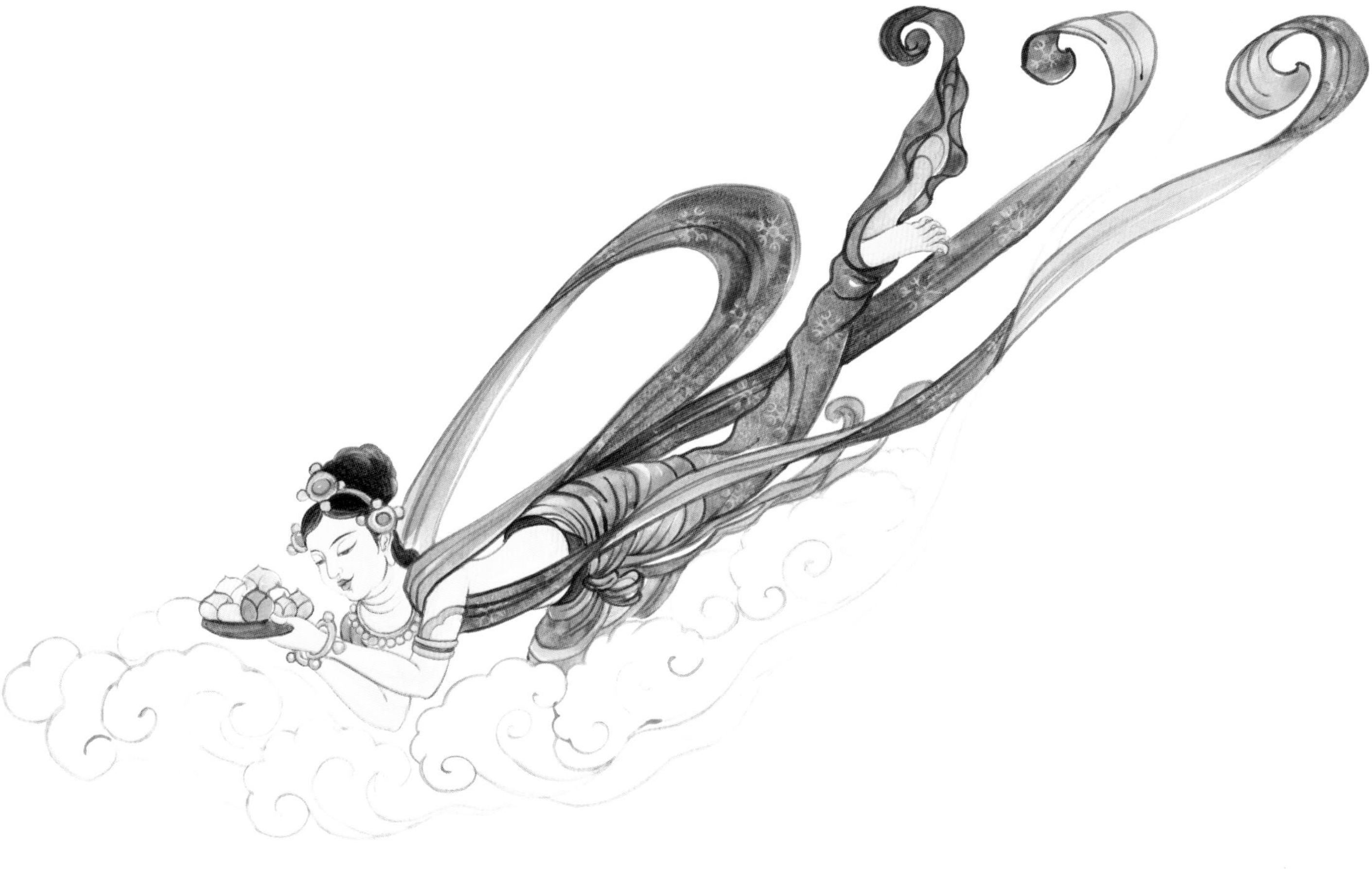

图：张春佳 Painted by: Zhang Chunjia

飞天服饰
莫高窟盛唐第39窟主室西壁
龛内

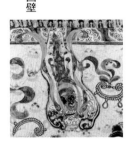

盛唐第39窟主室西壁龛内共五身飞天，本图典选取龛顶的一组三身，此图是中央的一身。由于整体布局大体呈左右对称状，因此该飞天的姿态，画家将其绘制为正面乘云向下飞的造型。两侧飘带、腰带翻卷的姿态也大体对称。这一组飞天之下是卧佛，飞天持盘托花蕾，并与飞扬的散花和璎珞供养涅槃佛。这样的气氛渲染可见《大般涅槃经》记载："诸天于空散曼陀罗华、摩诃曼陀罗华、曼殊沙华、摩诃曼殊沙华，并作天乐种种供养。"同时，众多佛经中对此类供养都有记载，也就是以天衣璎珞供养佛。飞天着头饰、颈饰、臂钏等饰物，这在古印度佛教石窟遗迹中多有此类形象出现。古印度佛教中天女的装饰往往非常华丽，周身上下饰品繁多，以饱满的肌体形态展示具有凡间女子理想形态特征的天神。这当然与印度的传统民族服饰习惯息息相关，及至中国，这些依然保留在壁画绘制中。同时由于唐代社会的文化包容力和亲和力，对于众多外来文化都持开放的态度，与本土进行融合时，在不同的领域进行着有选择的统一，莫高窟盛唐飞天展示出来的形象也是一种文化融合的趋势。

（文：张春佳）

There are five Flying Apsaras in the west niche of the main chamber in Cave 39 dated to the high Tang Dynasty. This book selects a group of three bodies from the top of the niche, and this painting is a body in the center. Because the overall layout is roughly symmetrical, the artist painted this one as flying down on clouds in the front. The rolling effect of ribbons and belts on both sides is also generally symmetrical. Below this group of flying Apsaras is reclining Buddha. The Flying Apsaras holds flower buds and offers Nirvana Buddhas with scattered flowers and keyūra. This atmosphere can be seen in Mahaparinirvana Sutra that "the celestial beings scatter Flos Daturae flowers, big Flos Daturae flowers, Manshusha flowers and big Manshusha flowers in the air, and make various heavenly music for offering." At the same time, such offerings are recorded in many Buddhist scriptures, that is, to provide Buddha with heavenly clothes and keyūra. The Flying Apsaras wears headdress, neck ornament, armlets and other ornaments, these also appeared in many ancient Indian Buddhist grottoes. In ancient Indian Buddhism, the decoration of heavenly women is often very gorgeous, there are many ornaments around the body, and the heavenly beings are displayed in the ideal body shape which desired by people. Of course, this is closely related to India's traditional national clothes, which are still preserved in mural painting in China. At the same time, due to the cultural tolerance and affinity of the society in the high Tang Dynasty, it held an open attitude towards many foreign cultures, when integrating with the local, it selectively unified in different fields. The images displayed in Flying Apsaras dated to the high Tang Dynasty at Mogao Grottoes are also manifested a trend of cultural integration.

(Written by: Zhang Chunjia)

图：张春佳 Painted by: Zhang Chunjia

　　第45窟是敦煌莫高窟盛唐时期的代表洞窟之一，为覆斗顶中小型窟，中唐、五代重修。主室窟顶画团花井心藻井，四周绘有垂角帷幔，四披画千佛。主室西壁开平顶敞口龛，龛内彩塑一铺七身。龛顶画菩提华盖，华盖上画法华经变见宝塔品，帐形多宝塔内二佛并坐，两侧环绕众多菩萨，整个龛顶华丽庄严。龛壁画火焰纹团花佛光，佛光两侧画菩萨各两身。龛外南北侧分别为中唐画观世音、地藏各一身，两侧存力士台墩各一。南壁盛唐画观音经变一铺，中央画观世音菩萨立像，两侧描绘了观世音菩萨救济诸难及三十三现身的场面。北壁盛唐画观无量寿经变一铺，中部画西方净土世界，东侧画未生怨，西侧画十六观。东壁门南盛唐画观世音菩萨一身，门北中唐画地藏、观世音各一身。

　　此窟塑像以精美写实而著名，为盛唐时期最具代表性的一铺彩塑。龛内塑像包含跌坐佛、阿难、迦叶、观世音、大势至及南、北天王共七身圆塑，并有序分布在敞口龛内。整窟塑像人物造型准确、姿态生动、个性鲜明，作者根据人物身份、性格、年龄，分别刻画每个人物的表情、身姿、衣饰，极具感染力地展现出佛陀的神圣庄严、迦叶的苦修沉稳、阿难的聪慧机敏、菩萨的温婉慈祥、天王的威武凶猛，显示出作者出神入化的技艺。塑像背后的龛壁和龛顶的壁画，与塑像绘塑结合，共同构成"净土说法"的场景。每尊塑像的服饰色彩、图案、搭配尤为巧妙，阿难的裙襦华丽富贵，迦叶的山水袈裟庄重沉着，菩萨的衣饰精致华美、色彩鲜艳，天王的甲胄层次分明、图案细腻，侧面体现出盛唐时期的纺织品种类繁多，配色艳丽，工艺精湛。

　　主室南壁的观音经变画面内容完整，线条流畅，色彩保存良好，榜题字迹清晰，堪称盛唐代表作。画面根据法华经变观音普门品，以观世音菩萨为主尊画于通壁壁画中央，两侧以山水和榜题间隔，描绘了观世音菩萨普度众生、救助世人的场景，其中胡商遇盗，航海遇难，大辟，求男得男、求女得女的画面反映出盛唐的世俗生活及衣冠制度。

（文：杨婧嫱）

Cave 45 is one of the representative caves in Dunhuang Mogao Grottoes of the high Tang Dynasty. It is a medium-sized cave with truncated pyramidal ceiling, which was repaired in the middle Tang and Five Dynasty. The cave top of the main chamber is painted with a round flower pattern as caisson center, surrounded by hanging valance and Thousand Buddhas. The west wall of the main chamber has a niche with a flat ceiling, in which has seven painted sculptures. The Bodhi canopy is painted on the top of the niche. Above the canopy is painted with Saddharmapundarika Sutra-Stuupadarsanparivartah. In the tent shaped treasured pagoda, two Buddhas sit side by side, surrounded by many Bodhisattvas on both sides. The whole niche top is gorgeous and solemn. The niche murals have flame pattern, round flower pattern Buddha light, and two painted Bodhisattvas on both sides of the Buddha light. The north and south sides outside the niche are painted Avalokitesvara and Ksitigarbharaja Bodhisattvas dated to the middle Tang Dynasty, and one Vīra platform on each side. On the south wall is Avalokitesvara Sutra illustration dated to the high Tang Dynasty, and a standing Avalokitesvara Bodhisattva is painted in the center. On both sides, the illustration of Avalokitesvara Bodhisattva's relieves and thirty-three transformations are depicted. The Amitayurdhyana Sutra illustration on the north wall dated to the high Tang Dynasty, the West Pure Land World is painted in the middle, Ajatasattu's story on the east side, and Sixteen Visualization is painted on the west side. Avalokitesvara Bodhisattva is painted on the south side of the east door dated to the high Tang Dynasty, and Ksitigarbharaja and Avalokitesvara are painted on the north side of the door dated to the middle Tang Dynasty.

This cave's sculptures are famous for their exquisite realism. They are the most representative painted sculptures of the high Tang Dynasty. The sculptures in the niche include seven sculptures of one sitting Buddha, Ananda, Kasyapa, Avalokitesvara, Mahasthamaprapta and the south and north Maharāja-devas, which are orderly distributed in the niche. The figures in the whole cave are accurate in shape, vivid in posture and distinctive in personality. According to the identity, character and age of the characters, the sculptor depicted the expression, posture and clothing of each character accordingly, the sacred solemnity of the Buddha, the hardship and calmness of Kasyapa, the intelligence and agility of Ananda, the kindness of Bodhisattva and the mighty and ferocious of the Maharāja-devas, showing the sculptor's superb skills. The niche wall behind the sculptures and the mural on the top of the niche together with the sculptures forming the scene of "Pure Land Dharma Assembly". The dress color, pattern and collocation of each sculpture are particularly ingenious. Ananda's robe is gorgeous and rich, Kasyapa's landscape kasaya is solemn and calm, the Bodhisattvas' clothes are exquisite, gorgeous and colorful, and the Maharāja-devas' armors have clear layers and exquisite patterns. All these reflect the wide variety of fabrics, gorgeous colors and exquisite workmanship in the high Tang Dynasty.

The painting of Avalokitesvara Sutra illustration on the south wall of the main chamber is complete, with smooth lines, well preserved colors, and clear handwriting inscriptions, which can be called the representative work of the high Tang Dynasty. The painting was done according to Saddharmapundarika Sutra-Samantamukhaparivartah. Avalokitesvara Bodhisattva is mainly painted in the center of the wall, and the landscape and inscriptions on both sides divided the whole picture into different sections to depict the scenes of Avalokitesvara helping and saving all sentient beings. Among them, the painting of Foreign Merchants Encounter Bandits, Marine Perils, Decapitation and Praying for Boy and Praying for Girl all reflect the secular life and dress habit of the high Tang Dynasty.

(Written by: Yang Jingqiang)

The Bodhisattva's clothes
on the south side of the west niche in the main
chamber of Cave 45 dated to the high Tang
Dynasty at Mogao Grottoes

菩萨服饰
莫高窟盛唐第45窟主室西壁
龛内南侧

盛唐第45窟主室西壁龛内南侧戴金冠菩萨，此菩萨的上身披络腋，下着花色缠腿薄纱阔裙，腰部有绣花腰裙，并垂缀长带，跣足轻踩莲花。

菩萨头上佩戴卷草纹镶嵌宝珠金冠（也称卷云冠），冠体的底部镶嵌有三颗莲花纹宝珠，其左右两侧有如意火焰纹饰，高高膨起的蔓草纹犹如卷云一般在头顶翻卷，其视觉效果颇为壮观。冠缯飘垂而纷披耳际，浓发披肩。菩萨面容圆润，修眉秀目，嘴角上扬，温婉柔美；四肢修长，玉手和秀足各有姿态。

在艺术表达上，菩萨造型静中有动，采用对比色配置，设色以平铺为主兼具晕染，用线粗细有序，平面与立体相呼应，别具艺术风格和个性化特色。

（文：刘元风）

In the west niche of the main chamber of Cave 45 dated to the high Tang Dynasty, there is a Bodhisattva on the south side who wears a golden crown on His head and Luoye on the upper body, a colorful tulle skirt on legs, an embroidered waist wrap around the waist, a long belt hanging, barefoot and stands on a lotus cushion lightly.

The Bodhisattva wears a golden crown with scrolling vine pattern inlaid with jewels(also known as the scrolling cloud crown), and the bottom of the crown inlaid with three lotus-pattern jewels and Ruyi flame pattern on both left and right sides. The high and bulging vine pattern scrolling on the top of His head like a scrolling cloud, and its visual effect is quite spectacular. The crown laces hang down and fall over the ears, and the thick hair cover the shoulders. The Bodhisattva's face is round, the eyebrows are trimmed, the corners of the mouth are raised, gentle and soft. The slender limbs, jade like hands and beautiful feet have their own features.

In terms of artistic expression, this Bodhisattva's modeling is dynamic within static. It adopts contrast color configuration, mainly flat coloring also used shading technique. Thin and thick lines are well managed, and the plane and three-dimensional echo with each other, which has unique artistic style and personalized characteristics.

(Written by: Liu Yuanfeng)

Venerable Ananda's
clothes
on the south side of the west niche
in Cave 45 dated to the high Tang
Dynasty at Mogao Grottoes

阿难尊者服饰
莫高窟盛唐第45窟主室西壁
龛内南侧

盛唐第45窟主室西壁龛内南侧的阿难（以"多闻第一"著称的阿难，曾跟随释迦牟尼四十五年听法布教）彩塑像，上身内着华丽的"V"形领右衽半袖偏衫，其领、袖处装饰以缠枝花纹的锦绣贴边，下身穿绿色的褶裙，其底摆处有条形缠枝花纹装饰，另有土红色的贴边，与外披袈裟热烈的土红色相呼应，整套服装既宽绰又不乏雅致。

阿难形象英俊秀朗，仪态闲适，双手相交置于腹前，身体稍向后倾，重心移至左脚，姿态潇洒，举止文雅，充分体现出年轻的阿难尊者聪敏颖慧、信心满满的精神状态，是盛唐开元前后佛像彩塑中的经典作品之一。

（文：刘元风）

On the south side of the west niche in the main chamber of Cave 45 dated to the high Tang Dynasty, there is a painted sculpture Ananda (Ananda, who is famous for "the most hearer" and had followed Sakyamuni for 45 years to listen the Dharma teaching). His upper body is decorated with a gorgeous "V" shaped collar half-sleeved right lapel shirt. The collar and sleeves are decorated with brocade welts with tangled vine pattern, and the lower body is dressed in a green pleated Qun with striped tangle vine pattern at the hem. In addition, the earth red welt echoes with kasaya color. With the earth red kasaya robe, the whole set of clothes is both gorgeous and elegant.

Ananda's image is handsome and relaxed. His hands are placed in front of His belly, the body leaning back slightly, and His weight on His left foot. His posture is natural and unrestrained, and His behavior is gentle. This sculpture fully reflects the young Ananda's intelligence and confidence. This is one of the classic works of Buddhist statue around Kaiyuan period of the high Tang Dynasty.

(Written by: Liu Yuanfeng)

Ananda's Sankaksika and skirt hem pattern
in the south side of the west niche in the main chamber of Cave 45 dated to the high Tang Dynasty at Mogao Grottoes

阿难尊者僧祇支和裙缘图案
莫高窟盛唐第45窟主室西壁龛内南侧

卷草纹是唐代壁画图案艺术的重要组成部分，呈"S"形波状茎蔓骨架，饰有花、枝、叶等装饰纹样。《汉语大词典》："海榴，即石榴，又名海石榴。"将石榴装饰在"S"形主藤蔓纹样上，称为石榴卷草纹。此服装边饰花枝叶蔓，丰厚饱满，让人联想到盛唐壁画中人物"丰腴赋体""曲眉丰颊"的艺术特征。

团花图案是莫高窟盛唐时期代表性纹样之一，该图案位于莫高窟第45窟西壁佛龛南侧弟子塑像裙摆处，整幅图案以二方连续半团花图案为主。半团花以团花变体性图案姿态出现在佛、菩萨、天王、力士等服饰上，体现了团花图案应用之广泛多样。

（文：姚志薇）

Scrolling vine pattern is an important part of the pattern art of murals of the high Tang Dynasty. It is an "S" shaped wavy vine frame, decorated with flowers, branches, leaves and other decorative elements. *Chinese Dictionary*: "Hailiu, namely pomegranate, also known as sea pomegranate." Pomegranate is decorated on the "S" shaped main frame, and the pattern is called Pomegranate Scrolling Vine Pattern. This clothes is richly decorated with flowers and leaves pattern, which reminds people of "plump body" and "curved eyebrows and plump cheeks" artistic characteristics in the high Tang Dynasty murals.

The round flower pattern was one of the representative patterns in the high Tang Dynasty at Mogao Grottoes. This pattern is located at the skirt hem of the disciple sculpture on the south side of the west niche of Cave 45 at Mogao Grottoes. The whole pattern is mainly two in a group repeated semi-round flower. Semi-round flower pattern appears in Buddha, Bodhisattva, Maharāja-devas, Vīra etc. clothes as round flower variant pattern, which reflects the wide and diverse application of round flower pattern.

(Written by: Yao Zhiwei)

图'' 姚志薇　Painted by: Yao Zhiwei

The Bodhisattva's clothes
on the north side of the west niche in the
main chamber of Cave 45 dated to the high
Tang Dynasty at Mogao Grottoes

菩萨服饰
莫高窟盛唐第45窟主室西壁
龛内北侧

　　盛唐第45窟主室西壁龛内北侧左胁侍菩萨，其面相丰满圆润，肌肤光洁细腻。上半身斜披络腋，下半身着轻薄的长裙，腰系锦绣罗裙，长裙和罗裙均装饰以缠枝连续性纹样。项链、臂钏、手镯金光璀璨，华丽夺目。菩萨头部云髻高耸，曲眉丰颊，双目微垂，嘴角上翘，神情恬静慈祥，呈"S"形的站姿，体态丰腴而娇娆。

　　在菩萨彩塑的整体艺术处理上，已突破了宗教艺术的审美范畴，其形象和形体的塑造，将宗教范式与唐代世俗女性形象有机地融为一体，特别是菩萨胸、腰、臀曲线的节奏感与律动美的塑造，无疑达到了盛唐时期彩塑艺术的最高水准。

（文：刘元风）

The attendant Bodhisattva on the left side of the west niche in the main chamber of Cave 45 dated to the high Tang Dynasty. His face is plump and round, and His skin is smooth and delicate. The upper body is wrapped with Luoye, the lower body is wearing a thin long skirt, and the waist is wrapped by a beautiful Luo skirt. Both the long skirt and Luo skirt are decorated with continuous tangled vine pattern. Necklace, armlets and bracelets are resplendent and gorgeous. This Bodhisattva has a tall hair bun, curved eyebrows and plump cheeks, slightly drooping eyes, upturned mouth corners, and a quiet and kind look, standing in an "S" shape posture, plump and charming.

In terms of the overall artistic treatment of this painted Bodhisattva sculpture, it has broken through the aesthetic category of religious art. The face and body shaping integrates religious paradigm and secular female aesthetic of the Tang Dynasty, especially the Bodhisattva's chest, waist and hip curves shaping, which undoubtedly reached the highest level of painted sculpture art during the high Tang Dynasty.

(Written by: Liu Yuanfeng)

The waist wrap and
skirt hem pattern of
the painted sculpture
Bodhisattva's clothes
in the west niche of the main chamber
in Cave 45 dated to the high Tang
Dynasty at Mogao Grottoes

彩塑菩萨
腰裙和裙
缘图案
莫高窟盛唐第
45窟主室西壁

该图案来自莫高窟第45窟菩萨塑像腰裙，上半部分以绿底为主色调，装饰红色五瓣小花，下半部分以间色菱形图案为主，两种图案组合，颜色冷暖交替、对比强烈又不失稳重和谐。

如意是佛教八宝之一，如意纹代表"回头即如意"的吉祥寓意。该图案位于莫高窟第45窟西壁佛龛南、北两侧菩萨裙身处，以如意纹结合团花图案为主，用色简洁明快，造型典雅舒展。

（文：姚志薇）

This pattern comes from the waist wrap of the painted sculpture Bodhisattva in Cave 45 of Mogao Grottoes. The upper part is mainly colored with green background and decorated with five-petaled red flowers. The lower part is mainly diamond pattern with different colors. The two patterns are combined with alternating cold and warm colors, which has strong contrast and harmony.

Ruyi is one of the eight Buddhist treasures. Ruyi pattern represents the auspicious meaning of "turning back is as one wishes". The pattern is located on the south and north sides Bodhisattva's skirts in the west niche of Cave 45 at Mogao Grottoes. It is mainly composed by Ruyi pattern and round flower pattern. The color is simple and bright, and the shape is elegant and stretched.

(Written by: Yao Zhiwei)

The painted sculpture
Bodhisattva's skirt
pattern
on the south side in the west
niche of the main chamber in Cave
45 dated to the high Tang Dynasty
at Mogao Grottoes

彩塑菩萨
长裙图案
莫高窟盛唐第
45窟主室西
壁龛内南侧

该图像位于莫高窟盛唐第45窟主室西壁龛内南侧菩萨长裙靠近腿的内侧位置处，以带状二方连续图案为主，主要的花朵图案中心以桃心二裂瓣为主，四片花叶包裹成圆形，其外部围绕四组三裂瓣的花朵呈"米"字形，周围饰小碎花，整幅图案零零整整，雅丽而时尚，搭配得宜，独具匠心。

（文：姚志薇）

This image is located at the inner side of the Bodhisattva's leg on the south side in the west niche of the main chamber in Cave 45 dated to the high Tang Dynasty at Mogao Grottoes. It is mainly a stripe-shaped two in a group repeated pattern. The main flower pattern center is mainly a heart shaped two lobes, and four petals surrounded in a circle. The four petals which have three lobes around are in a cross arrangement, surrounded by small broken flowers. The whole pattern is neat, elegant and fashionable, well designed.

(Written by: Yao Zhiwei)

图：姚志薇　Painted by: Yao Zhiwei

The painted sculpture
Maharāja-devas' clothes
on the north side of the west niche in the main
chamber of Cave 45 dated to the high Tang
Dynasty at Mogao Grottoes

彩塑天王服饰
莫高窟盛唐第45窟主室西壁
龛内北侧

盛唐第45窟主室西壁龛内北侧北方多闻天王彩塑像，其双眉紧锁，两目怒视，胡须飘散，左手叉腰，右手持兵器，身着甲胄，一派气冲霄汉的威武气概。

天王身着的是盛唐时期最典型的"金甲"，与初唐时期的"铁甲"不同，其头顶束髻，护领掩膊，兽头含臂，身甲、胸甲、髀裈、战裙、行縢、乌皮靴齐备。甲的形状有鳞形、长方形、六边形等，其中胸腹为重点保护部位，胸腹部的护甲以铜、铁金属为主，腹部的金属具有光亮，集保护性、功能性和审美性于一体。巧合的是，在同时期的欧洲罗马军中也有相连的战甲出现，由于作战的形式和兵器的不同，战甲的造型也各有特点。

（文：刘元风）

On the north side of the west niche in the main chamber of Cave 45 dated to the high Tang Dynasty, there is a painted sculpture Vaiśravana. His eyebrows are frown, eyes glare, beard scattered. His left hand is on His waist, His right hand holding a weapon, and He is dressed in armor, looks very muscular.

This Maharāja-devas is wearing the most typical "golden armor" during the high Tang Dynasty, which different from the "iron armor" in the early Tang Dynasty. He has a hair bun, a protective collar covering shoulders, beast heads around arms, and full body armor, chest armor, leg armor, battle skirt, leg bandage and black leather boots. The shape of armor pieces can be scaly, rectangular and hexagonal. The chest and abdomen are the key protection parts, mainly made of copper and iron metals, and the metal on abdomen is bright. This suit integrates protection, function and aesthetics all together. Coincidentally, there were also contemporary bind armor in the European Roman army. Due to the different combat situations and weapons, the armor shapes also have their own characteristics.

(Written by: Liu Yuanfeng)

The painted sculpture
Venerable Kasyapa's clothes
on the north side of the west niche in the main
chamber of Cave 45 dated to the high Tang
Dynasty at Mogao Grottoes

彩塑迦叶尊者服饰
莫高窟盛唐第45窟主室西壁
龛内北侧

迦叶是释迦牟尼的大弟子，原为婆罗门信徒，43岁时皈依佛门，以"苦修"著称。在第45窟主室西壁龛中，位于佛祖身旁的迦叶双眉紧蹙、目光炯炯，虽瘦骨嶙峋，但神态中透露出睿智与坚毅，俨然一位满腹经纶、思想深邃、通达世故、沉着老练的高僧形象。迦叶内着僧祇支，下穿百褶裙，外披田相袈裟，脚上穿履。僧祇支、褶裙、袈裟上均有图案及饰边装饰，衣饰图案描绘细腻、施色富丽。这身迦叶塑像的服装形态典雅庄重、衣纹流畅，人物造型生动传神。

迦叶所着的田相袈裟也称田相衣，因由条相将袈裟分成若干方块、形如水田而得名，也称水田袈裟。在敦煌莫高窟壁画与雕塑中，初唐以来出现的诸多佛弟子形象多穿田相袈裟。《释氏要览·法衣》说田相袈裟："象征田畦贮水，生长嘉苗，以养形命，法衣之田，润以四利之水，增其三善之苗，以养法身慧命也。"

（文：李迎军）

Kasyapa was the prime disciple of Sakyamuni, who was a Brahman and converted to Buddha at the age of 43, famous for "Ascetic Practice". In the west niche of Cave 45, Kasyapa, the one beside Buddha, has tight eyebrows and bright eyes. Although He is skinny, His expression reveals wisdom and perseverance, just like a well-educated, thoughtful, worldly, calm and experienced monk. Kasyapa wears Sankaksika inside, a pleated skirt on the lower body, covered with Field Pattern kasaya on the outside, and feet in shoes. There are patterns and decorative edges on Sankaksika, pleated skirt and kasaya. The clothes patterns are exquisite and richly colored, and the clothes form is elegant and solemn, the clothes lines are smooth, and the figure modeling is vivid and lively.

The Field Pattern kasaya worn by Kasyapa is also known as Tianxiangyi. It is named because the kasaya is divided into several squares by strips and looks like a paddy field. It is also known as the paddy field kasaya. In the murals and sculptures of Mogao Grottoes in Dunhuang, many Buddhist disciples since the early Tang Dynasty have mostly worn Field Pattern kasaya. *Buddhist Essentials · Dharma clothes* says about Field Pattern kasaya: "Just like water in field cultivate fine seedlings, the field on Dharma clothes, symbolize the water of four benefits, to nurture seedlings of three virtues, to cultivate the Dharma body and wisdom and life."

(Written by: Li Yingjun)

图：李迎军　Painted by: Li Yingjun

The celestial
woman's clothes
on the south wall in the main chamber
of Cave 45 dated to the high Tang
Dynasty at Mogao Grottoes

天女服饰

天女服饰

莫高窟盛唐第45窟主室南壁

此图位于盛唐第45窟主室南壁法华经变观音普门品东侧，表现的是大自在天作天女形象说法，一男子立地合十听法。左上侧榜题云："应以大自在天身得度者即，现大自在天身而为说法。"

天女头梳惊鹄髻，神态安详端庄，穿着襦裙式袿衣，袖型为"垂胡袖"。此种袖型袖管极宽大，至袖祛处急收，如黄牛喉下垂着的那块肉皱，文献记载其名为"胡"。天女的袖祛和领缘均为青金石色。天女的肩部有红色云肩，蔽膝与长裙间飘出几片间色华"髾"，形如刀圭状，足蹬尖头翘头履。画匠用艺术的手法将云肩、肘部缘饰和髾随风向后绘制，天女仿佛迎风而降，飞襳垂髾，栩栩然似燕尾垂曳，仙乐飘飘。

（文：董昳云、吴波）

This painting is located on the east side of the Saddharmapundarika Sutra-Samantamukhaparivartah on the south wall in the main chamber of Cave 45 dated to the high Tang Dynasty. It shows the image of Maheśvara transformed into a celestial woman who is giving a Dharma teaching. A man standing in front with palms folded and listening. The inscription on the upper left says: "If you need Maheśvara to save you out of suffering, the Bodhisattva will transform into Maheśvara to save you."

The celestial woman has a Startled Swan hair bun and looks serene and dignified, who wears Ru skirt style Guiyi gown, the sleeve type is "drooping foreign sleeve". This kind of sleeve is very wide but narrow at cuffs, just like the meat wrinkle hanging at yellow cow's throat, and the sleeve style was recorded in literature that the name is "Hu". The cuffs and collar edge are lapis lazuli, and there are red cloud shoulder covers on the shoulders, and several pieces of colored streamers float between Bixi and the long skirt, which are shaped like knifes, and Her feet in pointed upturned head shoes. The painter used artistic techniques to draw the cloud shoulder covers, elbow trims and streamers backward which seem like this celestial woman is descending against wind with flying streamers and ribbons. This is lifelike, like a swallow tail, with fairy music.

(Written by: Dong Yiyun, Wu Bo)

The Vaiśravaṇa's clothes on the east side of the south wall in the main chamber of Cave 45 dated to the high Tang Dynasty at Mogao Grottoes

毗沙门神服饰

莫高窟盛唐第45窟主室南壁

东侧

此图位于盛唐第45窟主室南壁法华经变观音普门品东侧角落，表现的是毗沙门神立地说法，一男子单膝跪地合十听法。右上侧榜题云："应以毗沙门神得度者即，现毗沙门神而为说法。"

毗沙门神头梳菩萨髻，身着石绿色戎装，盆领式护项与披膊相连，外套裲裆甲由皮襻连接前后衣片。胸下位置由螣蛇系扎，裲裆和螣蛇是相搭配的一种服饰。《旧唐书·舆服志》记载武官常服，如遇仪仗场合，"至于大仗陪立，五品以上及亲侍加裲裆螣蛇"。又《舆服志》释："螣蛇之制：以锦为表，长八尺，中实以绵，象蛇形。"毗沙门神下身穿着腿裙，腿裙之下有白色的衬裙。小腿位置配备胫甲，足蹬乌皮靴。小臂穿着护臂，在手肘处有一圈波浪状的袖缘饰，应为穿着于披膊内的半袖延续出的一圈白色袖缘饰，同类袖型可见于西魏第285窟四天王和龟兹石窟群壁画天王身上，推测为西域传入的一种袖型。此波浪袖也出现在同窟其他人物的服装上，为此窟特有风格。毗沙门神右手托窣堵波状宝塔，左手似在教导身旁的男子，一身戎装、立地说法、威风凛凛、气宇轩昂。

（文：董昳云、吴波）

This painting is located on the east corner of the Saddharmapundarika Sutra-Samantamukhaparivartah illustration on the south wall in the main chamber of Cave 45 dated to the high Tang Dynasty. It shows the standing Vaiśravaṇa giving Dharma teaching, a man kneels on one knee with two plams folded and listens to the Dharma. The inscription on the upper right says: "If you need Vaiśravaṇa to save you out of suffering, the Bodhisattva will transform into Vaiśravaṇa to give you Dharma teaching".

The Vaiśravaṇa has a Bodhisattva hair bun and a malachite green uniform. The basin shaped neck armor is connected with arm covers, and the front and rear pieces of Liangdang armor of the outer coat are connected by leather belts. The lower part of the chest is tied by Tengshe (螣蛇). Liangdang and Tengshe always appeared together, according to *Old Tang Book · Yu Fu Zhi* recorded the military officers daily clothes. In case of honor guard occasions, "As for big ceremony, officers Rank five up and close body guards have to add Liangdang and Tengshe on clothes". *Yu Fu Zhi* also explains: "Tengshe is made by brocade as cover, eight Chi long, silk in the middle, like a snake." Vaiśravaṇa wears battle skirt with white skirt under it. The lower legs are equipped with shin armor and black leather boots. The forearms wear forearm armor, and there are wavy sleeve edges at the elbows, which should be a circle of white cuffs from the half-sleeved shirt worn under the arm covers. The similar sleeve type can be seen on the four Mahārāja-devas in Cave 285 dated to the Western Wei Dynasty and the painted Mahārāja-devas in Qiuci Grottoes, which is speculated to be a sleeve type introduced from the Western Regions. This wavy sleeves also appear on clothes of other figures in the same cave, which is a unique style of this cave. Vaiśravaṇa holds a stupa shaped tower in the right hand, and the left hand is like giving teachings to the man beside Him. Vaiśravaṇa is dressed in military uniform and stands tall and dignified.

(Written by: Dong Yiyun, Wu Bo)

Prince Ajatashatru's clothes
on the east side of the north wall in Cave 45 dated
to the high Tang Dynasty at Mogao Grottoes

阿阇世太子服饰
莫高窟盛唐第45窟主室
北壁东侧

此图位于盛唐第45窟主室北壁观无量寿经变东侧，表现了《观无量寿经》里"未生怨"的故事。据佛经记载："时有一臣，名曰月光，聪明多智，及与耆婆，为王作礼。白言：……时二大臣说此语，竟以手按剑，却行而退。"从故事全图来看，北壁下部残损，故事通常是由下而上，此图下部画面的情节是有人将王后看望国王之事报告阿阇世太子。

此图则表现阿阇世王子拔剑欲弒王后，于是月光与耆婆谏劝阿阇世。该身复原男子为阿阇世太子，站立于台阶上，左手执剑，右手上扬，面对王后。阿阇世太子头戴通天冠（据《唐六典》卷十一记载："通天冠，加金博山，附蝉十二首，施珠翠，黑介帻，发缨翠绥，玉若犀簪导。"），是天子参与诸典礼时所穿戴之首服。阿阇世太子身上穿着的是右衽白练襦裙，束腰带，足蹬岐头履。

（文：董昳云、吴波）

This picture is located in the east side of the Amitayurdhyana Sutra illustration on the north wall of the main chamber in Cave 45 dated to the high Tang Dynasty, depicting the Ajatasattu's story from Amitayurdhyana Sutra. According to the Buddhist Scripture, "At that time, there was a minister named Yueguang, who was smart and wise, together with Qipo, bowed to the king. He said: ... When the two ministers said this, they put their hands on sword and walked away." From the whole picture we can see that the lower part of the north wall is damaged, and the story was usually depicted from bottom to top, and the plot in the lower picture of this story should be someone reported the queen's visit to Prince Ajatashatru.

This picture shows that Prince Ajatashatru, holding his sword and ready to kill the queen, so Yueguang and the Qipo tried to stop Ajatashatru. The man in the picture is Prince Ajatashatru, who stands on the platform, holding a sword in his left hand and rising his right hand, facing the queen. Prince Ajatashatru wears a Tongtian crown. According to volume 11 of *Tang Liu Dian*, "Tongtian crown, which has mountain shaped decoration in front, attached with twelve cicadas, with jade beads, black hair wrap, tied with green band and jade like rhinoceros horn hairpin." It is the crown worn by emperor when participating in various ceremonies. Prince Ajatashatru is wearing a right lapel white silk Ru skirt, tied a belt and feet in split heads shoes.

(Written by: Dong Yiyun, Wu Bo)

The minister's clothes
on the east side of the north wall in Cave 45
dated to the high Tang Dynasty at Mogao
Grottoes

大臣服饰
莫高窟盛唐第45
窟主室北壁东侧

此图位于盛唐第45窟主室北壁观无量寿经变东侧，表现了《观无量寿经》里"未生怨"的故事，与台阶上的阿阇世太子位于同一画面中，而该人物位于台阶下，按佛经内容，此位持剑进谏的应为臣子月光或耆婆。大臣头冠前后呈折角，似为进贤冠（《新唐书·舆服志》："进贤冠者，文官朝参三老五更之服也。"），以冠梁多寡区分等级高低。大臣身穿白练襦裙，肘部袖子呈现放射的波浪状，具有御风飞翔的意境，此袖子的形态与该窟其他人物一致，体现了画匠独特的艺术表现手法。

（文：董昳云、吴波）

This figure is located in the east side of the Amitayurdhyana Sutra illustration on the north wall of the main chamber in Cave 45 dated to the high Tang Dynasty, depicting the Ajatasattu's story from the Amitayurdhyana Sutra. He is in the same picture with Prince Ajatashatru, beside the steps. According to Buddhist Sutra, this person holding a sword should be minister Yueguang or Qipo. The minister's head crown has an angle shaped design at the back, which seems to be the Jin Xian crown. According to *the New Tang book·Yu Fu Zhi*: "The Jin Xian crown is the hat worn by civil officials and honorable official Sanlao and Wugeng." The official grade is distinguished by the number of crown beams. The minister is wearing a white silk Ru skirt, and the cuff around the elbow presents in a radiating wave shape, with the artistic conception of flying against the wind. The cuff design is consistent with other figures in the cave, which is an unique artistic expression of the painter.

(Written by: Dong Yiyun, Wu Bo)

图⋯ 吴波　Painted by: Wu Bo

世俗男女服饰

莫高窟盛唐第45窟主室南壁

西侧

此图为盛唐第45窟主室南壁西侧观音经变最左侧中上位置的一个场景。此经变以观世音菩萨为主尊，左右两侧下部画求诸苦难，上部画离"三毒"、得"二求"以及"三十三现身"等情节，各个情节之间以山与树间隔。

该场景描绘了恋爱中的男女。图中男子面带微笑，手持扇或诗书正与女子表白，女子娇媚含情回视，令人联想到唐传奇《会真记》中的场景。青年男子戴软脚幞头，着赭红色圆领袍服，束黑色革带，脚蹬乌皮靴，为盛唐男子常服。而女子梳倭坠髻，着襦裙，襦为蓝绿色、裙为赭红色，并搭饰浅色披巾，脚穿尖头软底鞋，为唐朝女子常服。初唐保留隋代修长而紧身的衣裙，到了盛唐由于女子以体态丰满为美，因此衣裙也随体型的变化显得宽松起来。图中女子体态丰满，衣裙符合盛唐时尚。

（文：赵茜、吴波）

This painting is in the west most middle upper part of the Samantamukhaparivartah illustration of the south wall in the main chamber of Cave 45 dated to the high Tang Dynasty. This sutra illustration has Avalokitesvara Bodhisattva as the main image in the center, the lower part of the left and right sides are painted all kinds of sufferings saved by Avalokitesvara, and the upper west part is painted with "away from the three poisons", "two wishes granted" and "thirty-three transformations", etc. Each plot is separated by mountains and trees.

The painting depicts a man and a woman in love. The man in the painting is smiling, holding a fan or poetry book, expressing his love to the woman. The woman is charming and looks back with affection, which is reminiscent of the scene in the legend of the Tang Dynasty *Hui Zhen Ji*. The young man wears soft ends Fu hat, ochre red round collar robe, black leather belt and black leather boots, which were often worn by men in the high Tang Dynasty. The woman combs a Wozhui hair bun and wears Ru skirt. The Ru is blue-green, the skirt is ochre red, and matched with a light colored silk scarf, feet in pointed soft sole shoes, which were often worn by women in the Tang Dynasty. In the early Tang Dynasty, the slender and tight clothes and skirts of the Sui Dynasty were retained. In the high Tang Dynasty, because women preferred plump as beauty, their clothes and skirts also became looser to fit the body shape. The woman in the painting is plump, and her clothes and skirts are in line with the fashion of the high Tang Dynasty.

(Written by: Zhao Xi, Wu Bo)

The man and
woman's clothes
on the west side of the south wall in the
main chamber of Cave 45 dated to the high
Tang Dynasty at Mogao Grottoes

世俗男女服饰

莫高窟盛唐第45窟主室南壁

西侧

第45窟主室南壁壁画内容为观音经变。在佛教中，观世音菩萨是西方极乐世界教主阿弥陀佛座下的上首菩萨，当众生有苦难时，只要称念他的名号，即可获得解脱苦厄。他还会就众生的因缘，化作种种不同的身份度化之。图中女子着襦裙装，上身为长衫，下着长裙，配披帛。裙子是唐代女子非常重视的下裳，制裙面料多为丝织品，裙色鲜艳，多为深红、绛紫、月青、草绿等。裙的式样用四幅连接缝合而成，上窄下宽，下垂至地，不施边缘。中唐以后，裙身越来越肥。披帛是受西域民族服饰影响演变而来的，披之于肩臂，在盛唐及五代最为盛行。襦裙装作为唐代女子常服中的重要代表，穿着轻便、时尚华丽，这与统治者思想政治开明是分不开的。民族与外来文化的大融合，让唐代襦裙区别于其他任何朝代的保守穿法，在中国古代服装史中，具有典型的时尚性。图中男子着唐代典型身长至足的襕袍，腰部用革带（蹀躞带）紧束，脚蹬皂靴。襕袍受胡服影响而成，其特点是在传统袍服下摆加一横襕，故而得名。南北朝时，北周武帝曾下令在袍、衫下加一横，以象征古代下裳。唐中书令马周曾上议在袍衫上加襕。这些都是为了仿古制。由此我们可以看出，在讲究礼制、规矩的传统文化影响下，中国服饰有着极强的传承性，同时，两种文化交融为新服饰的诞生提供了契机。

（文：王丽）

The mural on the west side of the south wall in Cave 45 shows Avalokitesvara Sutra illustration. In Buddhism, Avalokitesvara Bodhisattva is the prime Bodhisattva under the seat of Amitabha, the lord of the Western Pure Land. When any sentient being has suffering, one can get relief by saying His name. He will also help sentient beings in different identities according to the circumstances. The woman in the painting is wearing Ru skirt, with a long shirt on her upper body and a long skirt on the lower body, matched with silk scarf. Skirt was the lower clothes that women in the high Tang Dynasty favored, most of the skirt materials were silk, with bright colors, such as crimson, crimson purple, moon green, grass green, etc. The style of the skirt is four pieces of cloth stitched together, narrow at the top and wide at the bottom, hanging to the ground without trim. After the middle Tang Dynasty, the skirt became wider and wider. Silk scarf was introduced under the influence of ethnic clothes in the Western Regions, it was worn on the shoulders and arms and was most popular in the high Tang Dynasty and the Five Dynasty. As an important representative of women's regular clothes in the Tang Dynasty, Ru skirt is light and fashionable, which is inseparable from the ruler's ideological and political openness. The great integration of national and foreign culture makes the Ru skirt of the Tang Dynasty different from the conservative wearing way of any other dynasty. It has a typical fashion in the history of ancient Chinese clothing. The man in the painting is wearing a full-length robe typical in the Tang Dynasty, leather belt at waist (Diexie belt) and black boots on feet. Lan robe was influenced by foreign clothes, it is characterized by adding a piece of cloth below the hem of traditional robes, so it is named. In the Northern and Southern Dynasties, Emperor Wu of the Northern Zhou Dynasty ordered to add a piece of cloth below robes and shirts to symbolize ancient lower clothes. The head of the secretariat of the Tang Dynasty, Ma Zhou, proposed to add a piece of cloth below robes and shirts. These are all made to imitate ancient clothes, from this, we can see that Chinese payed great attention to etiquette and rules, under the influence of traditional culture, Chinese clothing has a strong inheritance tendency. At the same time, the integration of the two cultures provides an opportunity for the birth of new clothing.

(Written by: Wang Li)

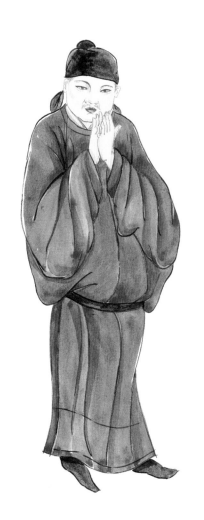

图'' 王丽   Painted by: Wang Li

The man's clothes
on the south wall in the main chamber of
Cave 45 dated to the high Tang Dynasty
at Mogao Grottoes

男子服饰

莫高窟盛唐第45窟主室南壁

《胡商遇盗图》中的这身正在打劫的盗贼头裹黑色幞头，幞头后边的长脚折系在脑后，身穿窄袖圆领袍，袍长至膝盖位置且袍的两侧有开衩，开衩处露出内衬的浅绿色衬袍，腰间系革带，腿上缠行缠，脚穿麻线鞋。

行缠是长带状的缠腿布，使用时自膝盖缠裹至脚踝，由于行缠裹紧了裤口，使小腿部分的服饰造型简洁利落而更便于活动，因此在劳动人民中被普遍使用。麻线鞋同样是盛唐劳动者普遍使用的服饰品，鞋体以麻线编结而成，吐鲁番阿斯塔那第106号墓曾出土唐代的麻线鞋实物，与壁画中绘制的造型基本一致。

主室南壁西侧描绘砍头场景的这组人物动态生动、惟妙惟肖，画面共有四人，其中两人拉绳、一人挥刀欲砍、一人待行刑。画面左侧的人物头裹黑色幞头，幞头外裹红色头巾，身穿白色窄袖圆领袍，袍长至膝盖位置，腰间系革带。

壁画上腿与鞋的形象只余下棕色剪影，根据画面中其他人物着装形象推断，此人应在腿上缠裹行缠，脚穿麻线鞋。

（文：李迎军）

In the painting of *Foreign Merchants Encounter Bandits*, the robber's head is wrapped in black Fu hat, the long ends behind the Fu hat are tied behind his head, and he wears a narrow-sleeved round collar robe, which is knee long and has slits on both sides. The slits revealed a light green under robe with leather belt around waist, the legs are wrapped by Xingchan (行缠), and feet in hemp fiber shoes.

Xingchan is a long strip of cloth, which wrap from knee to ankle when in use. Because the puttees tighten the trouser cuffs, so the lower legs look neat and more convenient for activities. Therefore, it is widely used among the working people. Hemp fiber shoes were also commonly used by workers in the high Tang Dynasty. The shoe body is woven by hemp fibers. The hemp fiber shoes unearthed from No. 106 tomb in Astana, Turpan dated to the Tang Dynasty which is basically consistent with the shape drawn in the murals.

This group of people in the beheading scene on the west side of the south wall in the main chamber is realistic and vivid. There are four people in the painting, including two people pulling ropes, one waving a knife to chop and one waiting to be executed. The figure on the left side of the painting who wears a black Fu hat, which wrapped in red kerchief. He wears a white narrow-sleeved round collar robe, which is knee long and tied with a leather belt around the waist.

The image of legs and shoes in the mural is only brown shape. According to the dress image of other characters in the painting, this person should have Xingchan around legs and wears hemp fiber shoes.

(Written by: Li Yingjun)

图'' 李迎军 Painted by: Li Yingjun

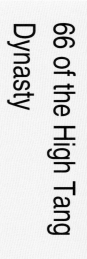

Dunhuang Mogao Grottoes Cave
66 of the High Tang
Dynasty

敦煌莫高窟盛唐

第66窟

　　第66窟为盛唐时期创建的覆斗顶形窟。主室窟顶藻井为云纹团花井心，四披画千佛。主室西壁开平顶敞口龛，龛内彩塑一佛、二弟子、二菩萨、二天王，经清代重绘。龛顶绘莲花宝盖及飞天等。龛壁浮塑火焰纹佛光，两侧共画八弟子、四菩萨、八伎乐。龛外南侧执连菩萨头戴化佛冠，北侧画观世音菩萨并配有榜题"救苦观世音菩萨"。龛内外绘塑结合共同组成十大弟子、八大菩萨的说法场面。主室南壁画西方净土变一铺，主室北壁中央画观无量寿经变一铺，东侧画未生怨，西侧画十六观。主室东壁损毁，仅门北存千佛数身。

　　此窟龛外北侧观世音菩萨为盛唐观音画像代表作之一。观世音菩萨，取用眼观察觉闻众生一心，而来救济世间悲苦，使之得解脱之意。在《观无量寿经》中，观世音菩萨与阿弥陀佛、大势至菩萨共称为"西方三圣"，阿弥陀佛作为主尊、观世音菩萨与大势至菩萨分别为左胁侍及右胁侍。随着净土信仰的盛行，观世音菩萨不仅作为"西方三圣"出现在说法图和经变画中，还出现大量和第66窟一样的形式，即出现在佛龛外侧两壁。此身观世音菩萨服饰图案精美，花纹繁复，色彩鲜艳，从侧面反映出盛唐染织技术高超，也体现出画师卓越的艺术才能。

（文：杨婧嫱）

Cave 66 is a truncated pyramidal ceiling cave built in the high Tang Dynasty. The caisson on the top of the main chamber has cloud round flower pattern in the center, with four slopes painted with Thousand Buddhas. The west wall of the main chamber has a flat ceiling niche. In the niche, there are painted sculptures of one Buddha, two disciples, two Bodhisattvas and two Mahārāja-devas, which have been repainted in the Qing Dynasty. The top of the niche is painted with lotus canopy and Flying Apsaras, and the niche walls have flame pattern Buddha light relief and eight disciples, four Bodhisattvas and eight musicians painted on both sides. Outside the niche, a Bodhisattva with Buddha image crown holding a lotus on the south, Avalokitesvara Bodhisattva on the north which has inscription says "suffering relieving Avalokitesvara Bodhisattva". The inside and outside the niche are a combination of paintings and sculptures to form the scene of ten disciples and eight Bodhisattvas in a Dharma assembly. In the main chamber, the Western Pure Land illustration is painted on the south wall, Amitayurdhyana Sutra illustration on the center of the north wall, and the painting on the east side is Ajatasattu's story, and the painting on the west side is Sixteen Visualization. The east wall is damaged, only a few Thousand Buddhas paintings exist on the north side of the door.

Avalokitesvara Bodhisattva on the north outside the west niche is one of the representative portraits of Avalokitesvara of the high Tang Dynasty. Avalokitesvara Bodhisattva uses His eyes to observe and hear all sentient beings and to relieve their sufferings and set them free. In Amitayurdhyana Sutra, Avalokitesvara Bodhisattva, Amitabha Buddha and Mahasthamaprapta Bodhisattva are called the "three saints of the west". Amitabha Buddha, as the Lord, Avalokitesvara Bodhisattva and Mahasthamaprapta Bodhisattva are left attendant and right attendant respectively. With the prevalence of Pure Land belief, Avalokitesvara Bodhisattva not only appeared in the illustration and Sutra painting as the "three saints of the west", but also appeared in a large number of forms like cave 66, which are on the two walls outside the niche. This Avalokitesvara Bodhisattva dress has exquisite patterns, complex lines and bright colors, which reflects the superb dyeing and weaving technology in the high Tang Dynasty and the artist's excellent artistic talent.

(Written by: Yang Jingqiang)

The Avalokitesvara Bodhisattva's clothes
on the north side of the west wall in the main
chamber of Cave 66 dated to the high Tang
Dynasty at Mogao Grottoes

观世音菩萨服饰
莫高窟盛唐第66窟主室西壁
龛外北侧

盛唐第66窟主室西壁龛外北侧观世音菩萨，面如满月，双目低垂，微微含笑，两耳垂肩，发辫飘散。菩萨扭腰出胯，身体重心落在右脚，立于莲花之上；左手下垂轻提璎珞珠串，右手轻拈胸前璎珞串珠。菩萨上半身着双色披帛，下半身穿薄纱花色透体长裙，腰间有锦绣罗裙，另配华丽的彩绦，冠带绕臂飘落。菩萨头戴摩尼宝珠冠，项链、臂钏、手镯、璎珞缀满全身。

在画面的艺术处理上，菩萨造型健美，身体向前微倾，更体现出与信众的亲近之感；色彩配置重彩叠晕，浓艳热烈；用线虚实相兼，生动传神，给人以极强的艺术感染力和视觉冲击力。

（文：刘元风）

On the north side of the west niche in the main chamber of Cave 66 dated to the high Tang Dynasty, the painted Avalokitesvara Bodhisattva's face looks like a full moon, His eyes looking down slightly, with little smile, long earlobes reaching to shoulders, windswept hair flying. This Bodhisattva twists His waist, the body weight on His right foot and stands on lotus. The left hand droops and gently lifts the keyūra beads, and the right hand gently twiddle the keyūra beads in front of the chest. The Bodhisattva wears two-color silk scarf on the upper body, and a long tulle skirt with flower pattern on the lower body, a brocade Luo skirt at the waist and colorful rope around, and the crown laces fall around the arms. A Cintamani crown on the head, with necklace, armlets, bracelets and keyūra all over the body.

In the artistic treatment of the picture, this Bodhisattva has a strong body shape and leaning forward slightly, which further reflects the sense of affinity to believers; used dark color and shading technique, warm and colorful; thick and thin lines all balanced, vivid and lively, giving a strong artistic appeal and visual impact.

(Written by: Liu Yuanfeng)

Avalokitesvara Bodhisattva's
Luoye and long skirt patterns
on the north side of the west wall in the main
chamber of Cave 66 dated to the high Tang
Dynasty at Mogao Grottoes

观世音菩萨络腋、
长裙图案
莫高窟盛唐第66窟主室西壁
龛外北侧

第66窟主室西壁龛外北侧观世音菩萨上身斜披土红色络腋，翻折处露出石绿色的内里，色彩对比鲜明，极具视觉冲击力。披帛上装饰着十瓣小朵花纹样，以浅石绿色为花瓣呈放射状展开，深绿色点缀花瓣边缘，鹅黄色的五个圆点簇在一起作为花心，纹样花地分明，灵动俏丽，错落有致。

观世音菩萨下身裹石绿色腰襷，内着腰裙和长裙。薄纱长裙以橙黄色为底色，四朵白色小花十字形排开，四周点缀绿色和蓝色的小叶片，形成对称的小簇花，色彩艳丽明快，纹样生动活泼。

（文：王可）

On the north side of the west wall in the main chamber of Cave 66, Avalokitesvara Bodhisattva's upper body is diagonally covered with earth red Luoye, and the folds revealed the malachite green inside, with bright color contrast and great visual impact. The silk scarf is decorated with ten-petaled small flower pattern, with light malachite green as the petals, spreading radially, and dark green as the edge of the petals, and five light yellow dots clustered together as the center of the flower. The pattern is clear, smart and beautiful.

Avalokitesvara Bodhisattva's lower body is wrapped in a malachite green waist wrap, with a short skirt and a long skirt inside. The tulle long skirt takes orange yellow as the background color, four small white flowers are arranged in a cross shape, surrounded by small green and blue leaves to form a symmetrical small cluster flower, with bright colors and lively patterns.

(Written by: Wang Ke)

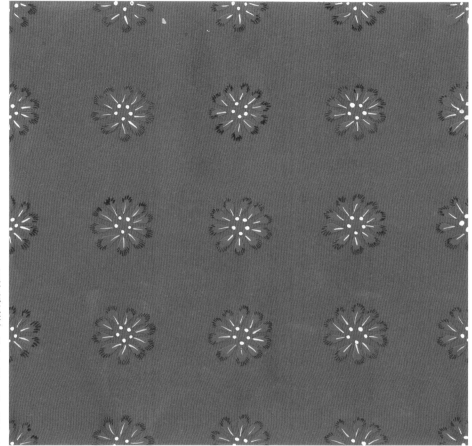

图'' 姚志薇　Painted by: Yao Zhiwei

图'' 王可　Painted by: Wang Ke

The attendant Bodhisattva's clothes on the north wall in the main chamber of Cave 66 dated to the high Tang Dynasty at Mogao Grottoes

胁侍菩萨服饰

莫高窟盛唐第66窟主室北壁

盛唐第66窟主室北壁胁侍菩萨，面容饱满，弯眉凤目，樱桃小口；顶束高髻，发辫披肩，佩戴火焰纹宝冠，冠体正中和两侧均镶嵌绿宝石。菩萨站姿呈"S"形，收腹提胯，跣足站于莲花之上，左手持莲花，右手持净瓶；上半身披红绿双色络腋，并在肩部打花结；下半身着花纹与条纹相兼的阔裙，腰系博带和彩绦，薄纱彩带环身飘散；冠带绕腕垂落，颈部的项链与璎珞装点其身，相互连接，随风而动。

在画面的艺术处理上，菩萨形体雍容典雅，用线洒脱挺拔，色彩为对比色配置，明快而醒目，呈现其丰富、多变、灵动的服饰艺术美感。

（文：刘元风）

The attendant Bodhisattva on the north wall in the main chamber of Cave 66 dated to the high Tang Dynasty has a round face, curved eyebrows and phoenix eyes, a small cherry mouth. A high hair bun tied on the top of His head, braided hair cover the shoulders, wearing a flame pattern crown, emeralds embedded in the middle and both sides of the crown. The Bodhisattva stands in an "S" posture, with His belly in and hips raised, barefoot standing on lotus. The left hand holding a lotus and the right hand holding a water vase. The upper part of the body is covered by red and green Luoye, and knotted at the shoulder. The lower part of the body is wearing a wide skirt with flower pattern and stripe pattern. The waist is tied with wide band and colorful rope, and the tulle color bands flying around the body. The crown laces fall around the wrists. The necklace and keyūra are connected with each other and move with wind.

In the artistic treatment of the picture, the Bodhisattva is graceful and elegant, with free and straight lines, contrast colors, bright and eye-catching, showing its rich, changeable and flexible aesthetic taste of clothes art.

(Written by: Liu Yuanfeng)

图：刘元风　　Painted by: Liu Yuanfeng

The offering
Bodhisattva's clothes
on the north wall in the main chamber of
Cave 66 dated to the high Tang Dynasty
at Mogao Grottoes

供养菩萨服饰
莫高窟盛唐第66窟主室北壁

　　盛唐第66窟主室北壁供养菩萨呈现为侧面形象，面相端庄、直鼻秀目，头束硕大而夸张的发髻，并向后、向下延展。供养菩萨挺胸收腹，玉树临风，跣足立于莲花之上，双手奉上宝石珠串。菩萨上身披络腋，肩上着红绿双色天衣（天衣正面为绿色，背面为红色），下身穿薄纱阔脚裤，另穿绿色的腰裙，长长的彩带随风飘扬；头上佩戴玉珠宝冠，冠体前面有三颗立体的宝石镶嵌，耳饰、项饰、腕饰及背部装饰尽情展现。

　　在艺术表现上，色彩采取了对比色的配置方式，用线刚劲流畅。供养菩萨的姿态优美，无疑是盛唐社会女性的真实写照，从中也反映出佛教的日益世俗化和本土化走向。

（文：刘元风）

The offering Bodhisattva on the north wall in the main chamber of Cave 66 dated to the high Tang Dynasty is presented as a profile, with dignified face, straight nose and beautiful eyes, and a huge and exaggerated hair bun on the head, which extends backward and downward. His shoulders back and belly in, like a jade tree in the wind. He stands barefoot on a lotus, and appreciate the gem beads in hands. The Bodhisattva's upper body is covered with Luoye, and wrapped a red-green shawl on shoulders (green on the front side and red on the back side). The lower body is wearing a wide-leg tulle pants, green apron, and long colored bands fluttering in the wind. The jade beads crown on His head has three jewels inlaid in front. The earrings, necklaces, bracelets and back decorations are displayed in details.

In terms of artistic expression, the color adopted contrast color configuration, and the lines are rigid and smooth. The graceful posture of this offering Bodhisattva is undoubtedly a true woman portrait of the high Tang Dynasty, which also reflects the increasing secularization and localization of Buddhism.

(Written by: Liu Yuanfeng)

图：刘元风　Painted by: Liu Yuanfeng

The attendant
Bodhisattva's clothes
on the west side of the south wall in the
main chamber of Cave 66 dated to the high
Tang Dynasty at Mogao Grottoes

胁侍菩萨服饰
莫高窟盛唐第66窟主室南壁
西侧

盛唐第66窟主室南壁胁侍菩萨，略施粉彩，眉间点画白毫，留有胡须，顶束高髻，发辫散落于肩。菩萨双手合十，结跏趺坐于莲花毯之上，身披络腋，条带绕臂飘落，下身穿花色长裙，蓝色丝带在腰部多层缠绕。

在绘画处理上，本尊胁侍菩萨的造型体现了盛唐人体美的新高度，即佛国人物无性别化的造型特征。而这种非男非女的艺术形象进一步显示出盛唐菩萨艺术的理性精神和审美倾向，这种美学特征通过画面的艺术表达手法而显现无疑。其华美的服饰，特别是富于律动感的天衣和飘带，典雅古朴的色彩，精致的装饰纹样，都充分彰显了盛唐造型艺术的人文气质和美学格调。

（文：刘元风）

The attendant Bodhisattva on the south wall in the main chamber of Cave 66 dated to the high Tang Dynasty, who put on make-up a little bit, with Urna between the eyebrows, bearded, high hair bun, and braids scattered on the shoulders. The Bodhisattva's two palms together, sit on the lotus cushion in a cross legged posture, covered by Luoye on the upper body, the silk scarf fall around arms, flower patterned skirt on the lower body, and wrapped blue bands around waist in multiple layers.

In terms of artistic treatment, the shape of this attendant Bodhisattva reflects the new height of body beauty during the high Tang Dynasty, that is, the asexual modeling characteristics of Buddhist figures. This non-male and non-female artistic image further shows the rational spirit and aesthetic tendency of Bodhisattva art during the high Tang Dynasty, through the artistic expression of picture, this aesthetic feature is undoubtedly manifested. His gorgeous clothes, especially the dynamic silk scarf and ribbons, elegant and simple colors and exquisite decorative patterns, fully demonstrated the humanistic temperament and aesthetic style of plastic arts during the high Tang Dynasty.

(Written by: Liu Yuanfeng)

图：刘元风　Painted by: Liu Yuanfeng

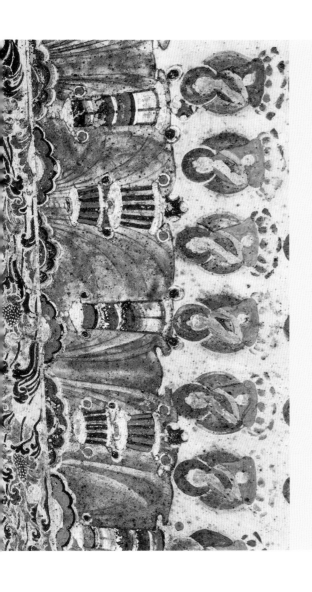

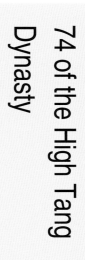

第74窟

敦煌莫高窟盛唐

Dunhuang Mogao Grottoes Cave
74 of the High Tang
Dynasty

第74窟是始建于盛唐时期的覆斗形顶窟。主室窟顶为莲花井心、四周垂幔璎珞的藻井，四披画千佛。西壁开斜顶敞口龛，龛顶画棋格帐顶图案，龛内彩塑已失，西壁浮塑佛光，佛光两侧各画二弟子，南北壁各画二弟子、二菩萨、一天王。龛外南北侧分别画地藏菩萨一身、毗卢舍那佛一身。南壁残存千佛，北壁画经变一铺。东壁门北存残存地藏及观世音各一部分。

此窟最精彩的部分为西壁的诸像，画师首先从人物构成和空间关系上就颇具巧思，将西、南、北壁共同组成八弟子、四菩萨、二天王庄严听法的场面。从绘画技法上，整铺壁画人物线条流畅、晕染柔和、层次丰富。从人物表现上，画师通过人物面孔轮廓、顾盼神情到身体姿态等细节刻画，生动地表现出菩萨的雍容智慧、弟子的沉稳持重、天王的气派威武，让人感觉人物形象生动、亲切真实，尽显画师艺术功底。此窟佛像、众弟子、地藏菩萨皆着袈裟，袈裟上画有缘边（四周边缘）与叶（中间横条和竖条）。这些条纹除了装饰外还起到固定的作用，从侧面反映了盛唐时期的袈裟形制。

（文：杨婧嫱）

Cave 74 is a truncated pyramidal ceiling cave built in the high Tang Dynasty. The top of the main chamber is a lotus in the center, surrounded by keyūra valance decoration caisson, the four slopes covered by Thousand Buddhas. The west wall has a niche with inclined top, and the niche ceiling is painted with flat checks pattern. The painted sculptures in the niche have lost, on the west niche wall has Buddha light relief. Two disciples are painted on both sides of the Buddha light, and two disciples, two Bodhisattvas and one Maharāja-deva are painted on each the north and south niche walls. The north and south sides outside the niche are painted with a Ksitigarbharaja Bodhisattva and a Vairocana Buddha. On the south wall partially has Thousand Buddhas, and the north wall has a sutra illustration. On the north side of the east door remains partially Ksitigarbharaja and Avalokitesvara Bodhisattva.

The most wonderful part of this cave is the images in the west wall. The painter had a lot of ingenuity in the composition of characters and spatial relations, and composed the west, south and north walls into a solemn scene of eight disciples, four Bodhisattvas and two Maharāja-devas listening to Dharma. In terms of painting techniques, the characters in the whole mural have smooth lines, soft halo shading and rich layers. In terms of character expression, the painter vividly shows the grace and wisdom of Bodhisattvas, the calmness and seriousness of the disciples and the majesty of the Maharāja-devas through the detailed depiction of the character's face outlines, eyes, expressions and body postures, which make people feel that the images are vivid, cordial and true, which fully reflect the artist's artistic skills. The Buddha, disciples and Ksitigarbharaja Bodhisattva in this cave all wear kasayas. The kasayas have painted trims (clothes edges) and leaf (horizontal and vertical lines in middle). These stripes not only are decoration but also have fixing function, they reflect the shape of kasaya in the high Tang Dynasty.

(Written by: Yang Jingqiang)

The Bodhisattva's clothes
on the north side of the west niche in the
main chamber of Cave 74 dated to the high
Tang Dynasty at Mogao Grottoes

菩萨服饰
莫高窟盛唐第74窟主室西壁
龛内北侧

　　盛唐第74窟主室西壁龛内北侧菩萨，形体姿态优雅，五官秀丽，眉间点画白毫；膊腕如藕，玉手纤纤；头束高髻，佩戴正方形嵌珠宝冠，红色的冠缯过耳垂肩。菩萨身披赭红色的络腋，下穿落地长裙，腰间装饰以多彩的围腰，自腰部垂有浅棕色飘带，透明的丝带在身前环绕飘拂。菩萨娇美的形象宛如盛唐世间的美女一般，比例协调，没有了之前隋、初唐时期菩萨头大身小（约5头身长）的视觉效果，使之充满了浓郁的世俗社会生活及文化气息。

　　在绘制表达上，先有线稿，后逐一敷色，敷彩时略施明暗，线与面有机结合，使画面产生一定的主体效果，体现盛唐时期绘画艺术的中西融合与革新意识。

（文：刘元风）

The Bodhisattva on the north side of the west niche in the main chamber of Cave 74 dated to the high Tang Dynasty has elegant posture, beautiful facial features and Urna between eyebrows. His arms and wrists are like lotus root, jade like hands are slender. High hair bun, a square crown inlaid with jewelry on head, and the red crown laces tied over the earlobes reach to shoulders. This Bodhisattva wears ochre red Luoye, long skirt, and the waist is decorated with a colorful waist wrap, with light brown ribbons hanging from waist, and transparent ribbons flying around the body. This charming image of Bodhisattva is like a beautiful woman in the high Tang Dynasty, with good proportion. Without the visual effect of Bodhisattva's big head and small body (about 5 heads long body) in the Sui and early Tang Dynasties, here this one is full of strong secular social life and culture sense.

In terms of artistic expression, line draft first, then apply colors. When applying color, shading technique applied slightly, the lines and planes are combined carefully, so that the painting has a certain main body effect, reflecting the integration and innovation awareness of Chinese and western painting art during the high Tang Dynasty.

(Written by: Liu Yuanfeng)

The Vairocana Buddha's
clothes
on the north outside of the west niche in the
main chamber of Cave 74 dated to the high
Tang Dynasty at Mogao Grottoes

毗卢舍那佛服饰
莫高窟盛唐第74窟主室西壁
龛外北侧

绘于西壁龛外北侧壁面的这身毗卢舍那佛，是少见的双手全部下垂（并且手中不持物）的立佛造型。尽管佛像目前的肤色已经氧化变黑，但依稀可以分辨五官的形状与神态。这身毗卢舍那佛面庞丰圆，弯眉细目，嘴角含笑，神情宁静，法相慈和，内着绿色僧祇支，下穿绿色长裙，僧祇支与长裙上装饰有风格一致的饰边纹样，外披通肩式红色田相袈裟，典雅庄重，气度雍容。

莫高窟中，盛唐时期袈裟的穿着方式主要有右祖式与通肩式两种，通肩式指将袈裟自身前经左肩、左臂覆至身后，再从身后覆右肩、右臂披至胸前，再复搭到左臂上的穿着方式。这身体态挺拔、双臂对称下垂的毗卢舍那佛采用通肩的方式穿着袈裟，左右对称的双臂造型塑造了袈裟流畅且有韵律的"U"形衣褶，画师流畅的笔意使这身端庄的佛像平添了一份灵动的韵味。

（文：李迎军）

On the north outside of the west niche, this Vairocana Buddha is uncommon, with both hands drooping (and holding nothing in his hands). Although the skin color of the Buddha has oxidized and turned into black, it still can be distinguished roughly that the shape and expression of the facial features. The face of the Vairocana Buddha is plump, curved eyebrows and small eyes, the corners of the mouth are rose, and the expression is quiet, looks merciful. He wears green Sankaksika inside, and green skirt on the lower body. The Sankaksika and skirt have same style decorative patterns on trims, and the outside wears both-shoulders-covered red kasaya with field pattern, looks elegant, solemn and peaceful.

In Mogao Grottoes, kasaya in the high Tang Dynasty mainly have two dressing ways: the right-shoulder-bared style and the both-shoulders-covered style. The both-shoulders-covered style refers to the way in which the kasaya covers the body from front over the left shoulder and left arm to the back, then covers the right shoulder and right arm to the chest from behind, and then puts it on the left arm. This tall and straight Vairocana Buddha with symmetrical drooping arms wears the both-shoulders-covered style kasaya. The symmetrical arms shaped the smooth and rhythmic U-shaped pleats of the kasaya. The painter's smooth lines added vividness to the dignified Buddha image.

(Written by: Li Yingjun)

图：李迎军

Painted by: Li Yingjun

The Maharāja-devas' clothes
outside the north wall of the west niche in the
main chamber of Cave 74 dated to the high Tang
Dynasty at Mogao Grottoes

天王服饰
莫高窟盛唐第74窟主室西壁
龛内北侧

西龛南北两壁最外侧两身天王像的身高明显高于龛内的菩萨像与弟子像，两身天王在西壁龛内最外侧顶天立地，神态威严。

天王的职责是护法镇邪、守护天界，多戴胄着甲、手持兵器。这身绘于西龛北壁的天王赤发虬须，愤然怒目，左手托宝珠，右手持剑，戴宝冠，穿护项、披膊、明光甲、胫甲，系束甲带与皮腰带，穿战靴。其中作为胸甲的明光甲也称明光铠，主要造型特征是左右两胸前各有一个凸起的圆形护胸甲片。据说，这种胸甲的形制最早出现在古希腊，辗转经中亚传入中国，进而在南北朝烽火连天的战局中发挥了重要作用。及至唐时期，这种闪闪发光的金属胸甲逐步完成了本土化的演变，与中国甲衣上各部分结构完美融合，成为中国甲胄必不可少的一部分，并对后世甲胄的发展产生了深远的影响。

（文：李迎军）

The heights of the two Maharāja-devas on the outermost sides of the north and south walls of the west niche are significantly higher than that of the Bodhisattvas and disciples in the niche. The two Maharāja-devas stand tall outside the west niche, look serious and dignified.

The duty of Maharāja-deva is to protect the Dharma, suppress evil and guard the heaven, usually wears armor and holds weapons. The Maharāja-deva painted on the north wall of the west niche has red hair and bearded, looks angry and eyes wide open. He holds a jewel in His left hand, and a sword in His right hand, wears crown, neck armor, shoulder covers, body armor, shin armor, armor fasten belt, a leather waist belt, and battle boots. Mingguang armor, also known as Mingguang Kai, is characterized by a convex circular breastplate on each the left and right chest. It is said that the shape of this kind of breastplate first appeared in ancient Greece and was introduced into China through Central Asia, which played an important role in wars during the Northern and Southern Dynasties. Until the Tang Dynasty, this glittering metal breastplate gradually completed the localization transformation, perfectly integrated in the structure of Chinese armor, and became an indispensable part of Chinese armor, which had a far-reaching impact on the development of later armor.

(Written by: Li Yingjun)

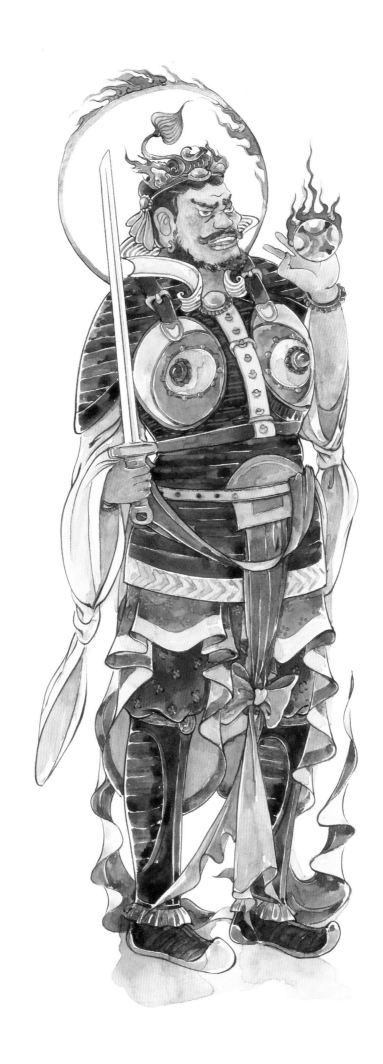

图：李迎军　Painted by: Li Yingjun

The Ksitigarbharaja
Bodhisattva's kasaya pattern
on the south outside the west niche in the main
chamber of Cave 74 dated to the high Tang
Dynasty at Mogao Grottoes

地藏菩萨袈裟图案
莫高窟盛唐第74窟主室西壁
龛外南侧

地藏菩萨在中国佛教中久负盛名，与观世音、文殊、普贤并称为"中国佛教四大菩萨"。第74窟主室西壁龛外南侧地藏菩萨现比丘相，作比丘装束，圆顶光头，身覆袈裟，手托摩尼宝珠，表示满足众生之愿望。这种形象的地藏菩萨，犹如一位云游四方的僧人，以救济世人之苦为己任，给人以亲切之感。外披的田相袈裟，坛（田相格）内为深褐色地，上有石绿色和石青色六瓣小朵花规律穿插排列。图案造型似梅花，花瓣浑圆，根部细窄，以平涂轮廓式方法表现。这种边缘清晰、造型简洁、色彩单一的纹样推测可用模板印花的工艺方式呈现。四周边缘与叶（中间的横条和竖条）上没有装饰，用沙黄色布条缝缀。此袈裟与敦煌莫高窟初唐第333窟主室西壁龛内弟子像袈裟类似，可见这种风格的袈裟由初唐一直流行到盛唐。

（文：王可）

Ksitigarbharaja Bodhisattva has long been famous in Chinese Buddhism. Together with Avalokitesvara, Manjusri and Samantabhadra, they are known as the four major Bodhisattvas in Chinese Buddhism. Outside the west niche of the main chamber in Cave 74, the Ksitigarbharaja Bodhisattva on the south side looks like a monk and is dressed in monk's clothes. His head is tonsured, the body is covered with kasaya, and the hands carry Cintamani representing to meet the needs of all sentient beings. This image of Ksitigarbharaja Bodhisattva is like a monk wandering around, holding the duty to relieve the sufferings of all sentient beings and gives people a sense of kindness. The kasaya He wears on the outside has field pattern, in the Tan (field pattern) is dark brown background with six-petaled small malachite green and azurite flowers interspersed regularly. The pattern looks like a plum blossom, with round petals and thin and narrow pedicels. They are expressed in flat coloring, so it is speculated that this pattern with clear edges, simple shapes and single color can be made by plate printing. The frames (the horizontal and vertical bars in the middle) are not decorated with pattern, instead sewn with sand yellow cloth strips. This kasaya is similar to the kasaya of the disciples in the west niche of the main chamber of Cave 333 dated to the early Tang Dynasty at Mogao Grottoes, Dunhuang. It can be seen that this style of kasaya has been popular from the early Tang Dynasty to the high Tang Dynasty.

(Written by: Wang Ke)

图" 王可　Painted by: Wang Ke

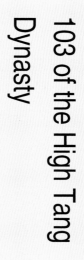

第103窟

敦煌莫高窟盛唐

Dunhuang Mogao Grottoes Cave
103 of the High Tang
Dynasty

第103窟是敦煌莫高窟盛唐时期的代表洞窟之一，为覆斗形窟。主室藻井为团花井心，四周画联珠、半团花、鳞纹、垂角幔帷，四披画千佛。主室西壁开平顶敞口龛，清修彩塑一佛、清重塑二弟子、清修二菩萨。龛外南北侧各绘一菩萨，各立一清塑菩萨于土台上。主室南壁绘法华经变一铺，画面中间绘序品，西侧绘化城喻品，东侧绘妙庄严王本事品。主室北壁绘观无量寿经变一铺，两侧分别绘未生怨及十六观。主室东壁绘维摩诘经变一铺，门南绘维摩诘，下绘方便品；门北绘文殊菩萨，下绘各国王子礼佛图；门上绘佛国品；三幅画面围绕窟门成"品"字排列。前室西壁存天王像。甬道南壁北壁分别为地藏菩萨和毗沙门天王各一身。

此窟壁画已经体现出典型的盛唐画风，是研究唐代艺术的重要资料。主室东壁的维摩诘经变尤为精彩。画师画技高超，用流畅有力的线条勾勒出画面中众人的神态，人物刻画生动写实，感染力极强。画师用焦墨勾线，略施淡彩，塑造出维摩诘"虬须云鬓，数尺飞动"的模样，其服饰衣纹清晰，衣褶流畅又富有变化，尽显著名画师吴道子为代表的唐代画风。南壁法华经变，山水层次深厚，色调青绿，意境隽永，也是盛唐山水画的代表作，体现出盛唐崇尚富丽的山水画风格。北壁城中有城，院中有院，楼阁金碧辉煌、结构清楚，反映出盛唐时期的建筑布局。

（文：杨婧嫱）

Cave 103 is one of the representative caves in Dunhuang Mogao Grottoes of the high Tang Dynasty. It is a truncated pyramidal ceiling cave, the center of the caisson in the main chamber is round flower pattern, surrounded by beads pattern, semi-round flower pattern, scale pattern, valance with triangle edges and Thousand Buddhas. The west wall has a flat ceiling niche, with painted sculptures but repaired in the Qing Dynasty of one Buddha, two disciples remade in the Qing Dynasty and two Bodhisattvas repaired in the Qing Dynasty. A Bodhisattva is painted each on the north and south sides outside the niche, and on each earth platform has one painted sculpture Bodhisattva made in the Qing Dynasty. The south wall of the main chamber is painted with Saddharmapundarika Sutra illustration, the Chapter of Ninaadaparivartah is painted in the middle of the painting, the Chapter of Prvayogaparivartah is painted in the west side, and the Chapter of Ptirvayogaparivartah is painted in the east side. On the north wall is painted with Amitayurdhyana Sutra illustration, and the two sides are painted with Ajatasattu's story and Sixteen Visualization respectively. The east wall is painted with the Vimalakīrti-nirdeśa Sutra illustration, the south side of the door is painted with the Vimalakīrti, and the lower part is painted with the Chapter of Skillful Means; Manjusri Bodhisattva is painted on the north side of the door, and the lower part is painted with princes of Various Countries Listening Dharma; above the door is painted with the Chapter of Buddhist World; the three paintings are arranged in the Chinese character structure "品" around the cave door. There is Maharāja-deva painting on the west wall of the antechamber. The south wall and the north wall of the corridor are painted with Ksitigarbharaja Bodhisattva and Vaiśravaṇa Maharāja-deva respectively.

The murals in this cave have reflected the typical painting style of the high Tang Dynasty and are important materials for the study of the high Tang Dynasty art. The Vimalakirti Sutra illustration on the east wall of the main chamber is particularly wonderful. The painter's painting skills are superb. He used smooth and powerful lines to show features and spirits of everyone in the painting. The characterization is vivid and realistic, with strong appeal. The painter used dark and dry ink to draw lines and applied few light colors to create the appearance of "long beard and hair flying in the wind" of Vimalakirti. His clothes have clear lines, smooth and varied pleats, showing the painting style of the high Tang Dynasty represented by the famous painter Wu Daozi. Saddharmapundarika Sutra illustration on the south wall, have complex landscape layers, green as the main color, and the artistic conception is deep. This is also a representative work of landscape painting of the high Tang Dynasty, reflecting the rich landscape painting style of the high Tang Dynasty. The north wall has city within city and courtyard within courtyard, and the pavilions are resplendent and the structure is clear, reflecting the architectural layout in the high Tang Dynasty.

(Written by: Yang Jingqiang)

Samantabhadra
Bodhisattva's clothes
on the south wall in the main chamber
of Cave 103 dated to the high Tang
Dynasty at Mogao Grottoes

普贤菩萨服饰
莫高窟盛唐第103窟主室南壁

盛唐第103窟主室南壁的普贤菩萨，面容端庄，蛾眉秀目，楚楚动人。双手作"普贤三昧耶印"，盘腿坐于莲花座之上。头束高髻，佩戴镶嵌蓝宝石的宝冠，冠体为红绿锯齿形纹饰，其上镶嵌双层7颗蓝宝石，正中上部的一颗体量最大，可见盛唐时期菩萨冠饰之华丽和精美。项链、臂钏、璎珞均镶嵌有蓝宝石。

菩萨上半身穿赭红色的络腋，下半身着土红色阔裙，裙摆镶嵌蓝色贴边，自冠缯垂下的长带绕肘飘洒。绿色薄纱天衣环绕腰部及身体左右。

在绘画艺术处理上，菩萨的造型生动，比例适度，画面用线或转折或挺括；色彩淡雅，清新明快，具有"落笔雄劲，敷色简淡"之吴道子的艺术风范。

（文：刘元风）

The Samantabhadra Bodhisattva on the south wall in the main chamber of Cave 103 dated to the high Tang Dynasty has a dignified face, beautiful moth eyebrows and moving eyes. Performing the "Samantabhadra Samadhi mudra" by both hands and sits cross legged on lotus seat. His hair is tied in a high hair bun and wears a treasure crown inlaid with sapphires. The crown body has red and green zigzag decoration which inlaid with seven sapphires in two layers. The upper one in the middle has the largest volume, which shows the gorgeous and exquisite crown decoration of Bodhisattva during the high Tang Dynasty. The necklace, armlets and keyūra are all inlaid with sapphire.

The Bodhisattva wears ochre red Luoye on the upper body and a broad earth red skirt on the lower body. The skirt hem has blue welt, and the long bands hanging from the crown fluttering around the elbows and a green gauze silk band around the waist and body.

In terms of artistic expression, the Bodhisattva's shape is vivid, the proportion is appropriate, and the lines are curved or straight; the colors are graceful, fresh and lively, which have Wu Daozi's artistic style of "vigorous lines and simple color".

(Written by: Liu Yuanfeng)

Vimalakirti's clothes
on the south side of the east wall in the
main chamber of Cave 103 dated to the
high Tang Dynasty at Mogao Grottoes

维摩诘服饰
莫高窟盛唐第103窟主室东
壁南侧

在盛唐第103窟主室东壁南侧维摩诘经变壁画中，维摩诘居士赤足斜坐高脚胡床之上，手持麈尾，凭几而坐。其上身内穿曲领中单，外着白色交领宽袍，最外层是褐红色披风。

维摩诘居士头裹白色纶巾，童颜鹤发，须髯飘洒，体现其满腹经纶、从容不迫、滔滔雄辩的鸿儒风范。

在绘画的艺术表现上，人物造型生动精确，形象刻画形神兼备，重点刻画人物眉宇、眼神及嘴角等处的精妙表达。壁画以刚柔相济的墨色线条，轻重对比的配色关系，以及"其傅彩于焦墨痕中，略施微染，自然超出缣素"的中原艺术范式，彰显出盛唐造型艺术的典型特征。

（文：刘元风）

In the painting of Vimalakirti Sutra illustration on the south side of the east wall in the main chamber of Cave 103 dated to the high Tang Dynasty, Vimalakirti lays on a high legged foreign bed with bare feet, holding his Zhuwei leaning on Ji. The upper body is wearing a curved collar Zhongdan, a white cross collar wide robe, and the outermost covers a maroon cloak.

Vimalakirti kulapati wrapped a white Guanjin on his head, white hair and a ruddy complexion, flying beard, these all reflecting his erudite, free and eloquent characteristic of a great master.

In the artistic expression of painting, the figure modeling is vivid and accurate, the image description has both form and spirit, and focused on the subtle expression of the figure's eyebrows, eyes and mouth corners. The ink lines are balanced with hardness and softness, the color matching relationship of light and dark contrast, and the Central Plains art paradigm of "the color is slightly painted between the dry ink lines, which naturally exceeds the painting itself" highlighting the typical characteristics of plastic art during the high Tang Dynasty.

(Written by: Liu Yuanfeng)

图：刘元风　Painted by: Liu Yuanfeng

此为盛唐第103窟主室东壁维摩诘经变中侍立于维摩诘帐旁的天女。第103窟绘于唐玄宗天宝年间，利用东壁窟门上部空间，用佛国品宝盖供养佛的故事将南北对立的"文殊问疾"和"维摩诘示疾"画面连接在一起。

天女右手执麈尾，左手拈花，欲用花戏弄舍利弗，出自观众生品中天女散花的典故。天女头梳惊鹄髻，用花束发髻，唐人喜爱将牡丹等鲜花簪于发髻之上，李白《宫中行乐词》曰："山花插宝髻。"万楚《茱萸女》有："插花向高髻"的诗句。天女整体穿着襦裙式袿衣，内着曲领中单小袖衫，外披右衽褐色大袖襦衫加半袖，领口处有曲折状缘饰，与盛唐第46窟南壁摩耶夫人彩塑像的领缘近似。腰间系"襳"，下着石绿色蔽膝和土黄色曳地长裙，下部有多片华"髾"飞舞，蔽膝向上翻滚，来表现天女"扬轻袿之猗靡兮"和"飘忽若神"之感。足蹬云头履。此身天宝年间的天女袿衣较初唐时期有所变化，蔽膝改为正面长而突出的围裹式，整体突出了天女灵动飘逸的神仙气质。

（文：董昳云、吴波）

The celestial woman's clothes beside Vimalakirti tent on the east wall of the main chamber in Cave 103 dated to the high Tang Dynasty at Mogao Grottoes

维摩诘帐旁天女服饰

莫高窟盛唐第103窟主室东壁

The celestial woman standing next to the Vimalakirti tent in Vimalakirti Sutra illustration on the east wall of the main chamber in Cave 103 dated to the high Tang Dynasty. Cave 103 was painted during the Tianbao period of Emperor Xuanzong of the Tang Dynasty, which used the upper space of the east door and the story of offering Buddha with canopy from the Chapter of the Buddhist World to connect the paintings of "Manjusri visitting Vimalakirti" and "Vimalakirti pretending sick" on both sides of the door.

The celestial woman holds Zhuwei in her right hand and twiddle the flower in her left hand. She wants to tease Sariputra by flowers, this story comes from the Chapter of Observing All sentiment Beings that the celestial woman scattering flowers to the audiences. The celestial woman has a Startled Swan hair bun and uses flower band to tie hair. The Tang Dynasty people liked to put peonies and other flowers in hair bun. Li Bai's *Gong Zhong Xing Yue Ci* says: "mountain flowers in treasure hair bun." Wan Chu's *Zhu Yu Nü* has: "arrange flowers in high hair bun". The celestial women wears Ru skirt style long clothes, with curved collar Zhongdan small-sleeved shirt inside and a brown large-sleeved Ru shirt with half-sleeved coat outside. There are zigzag trim at the neckline, which is similar to the collar of the painted sculpture of Mahamaya in the south wall of Cave 46 dated to the high Tang Dynasty. The waist is fastened by Xian, a malachite green Bixi and floor-length skirt on the lower body. At the lower part, there are many pieces of "Shao" flying and rolling upward to show the feeling of the celestial woman "flying lightly" and "wandering like spirit", and feet in cloud head shoes. Compared with the early Tang Dynasty, the celestial women clothes in the Tianbao period has changed, Bixi became longer and prominent into wrapping style, highlighting the free and elegant immortal temperament of the celestial women.

(Written by: Dong Yiyun, Wu Bo)

The celestial woman's clothes beside Manjusri Bodhisattva on the east wall of the main chamber in Cave 103 dated to the high Tang Dynasty at Mogao Grottoes

文殊菩萨身旁天女服饰

莫高窟盛唐第103窟主室东壁

此为盛唐第103窟主室东壁维摩诘经变中侍立于文殊菩萨身旁的天女。

天女梳惊鹄髻，戴花饰，整体服饰与维摩诘帐旁的天女基本相同，半袖袖口的一圈伞状缘饰更为明显，足蹬尖头翘头履。

（文：董昳云、吴波）

This is the celestial woman next to Manjusri Bodhisattva in Vimalakirti Sutra illustration on the east wall of the main chamber in Cave 103 dated to the high Tang Dynasty.

The celestial woman has a Startled Swan hair bun, wears a flower on head. The overall dress is basically the same as the celestial woman next to the Vimalakirti tent. The umbrella shaped trim decoration of the half-sleeved coat cuffs are more obvious, and her feet in pointed upturned heads shoes.

(Written by: Dong Yiyun, Wu Bo)

图：吴波　Painted by: Wu Bo

The Bodhisattva's clothes
on the south side of the east wall in the
main chamber of Cave 103 dated to the
high Tang Dynasty at Mogao Grottoes

菩萨服饰

莫高窟盛唐第103
窟主室东壁南侧

东壁的维摩诘经变以东壁门为参照，按"品"字形排列绘制了三部分内容，门上方为"佛国品"，门北侧为"文殊问疾"，门南侧为"维摩诘示疾"。"维摩诘示疾"画面下方描绘的是文殊与维摩诘辩论，因时间已晚，舍利弗腹中饥饿，维摩诘现神通力作化菩萨，借饭香积佛。这身送香饭的菩萨也因此被称作"香积菩萨"，画面中的香积菩萨胡跪在莲座上，双手托钵，向诸族王子、官属献香饭。菩萨面相丰满，云鬘高髻，宝冠束发，冠缯长垂，浓发披肩，上身着僧祇支，披帛垂绕，戴颈环、臂钏、手镯，下着红裙，系绿色围腰。整体造型姿态优美，形神兼备。

整铺壁画以线条的气韵生动著称，这身菩萨同样敷彩简淡，只有僧祇支上平涂了绿色。在绘画整理时，参照相关洞窟的菩萨形象为这身菩萨的服装适度着色。

（文：李迎军）

Vimalakirti Sutra illustration on the east wall can be divided into three parts in "品" layout. Above the door is "The Chapter of Buddhist World", the north side of the door is "Manjusri visitting Vimalakirti", and the south side of the door is "Vimalakirti pretending sick". Below the painting "Vimalakirti pretending sick" is Hua Bodhisattva, which means during the long debate between Manjusri and Vimalakirti, Sariputra was hungry, Vimalakirti used his magical power transformed into Hua Bodhisattva and went to Fragrance Buddha world and borrowed a bowl of rice. This Bodhisattva who delivers fragrant rice is also called "Xiangji Bodhisattva". In the painting, Xiangji Bodhisattva kneels by one knee on the lotus seat, holds the bowl by both hands and offers fragrant rice to princes and officials of all nationalities. The Bodhisattva has a plump face, a high hair bun, a treasure crown, tied hair, long crown laces and thick hair covers shoulders. The upper body wears Sankaksika, draped with silk scarf, necklace, armlets and bracelets, and red skirt and a green waist wrap on the lower body. The overall shape is beautiful both in form and spirit.

The whole mural is famous for the vivid lines. The Bodhisattva colored simple and light. Only the Sankaksika is green flat coloring. When doing the illustration, the Bodhisattva's clothes were moderately colored with reference to the Bodhisattva image of the relevant caves.

(Written by: Li Yingjun)

The fan-holding
servant's clothes
on the north side of the east wall
in the main chamber of Cave 103
dated to the high Tang Dynasty at
Mogao Grottoes

持扇侍从服饰

莫高窟盛唐第103窟
主室东壁北侧

在维摩诘经变画中，持扇侍从与维摩诘、文殊师利、舍利弗、菩萨、帝王、群臣、各国王子相比完全是一位"小人物"，但这位"小人物"却是各时期维摩诘经变画中频繁出现在帝王身边的形象，莫高窟第103窟主室东壁这身持扇侍从虽然敷彩简淡，却造型生动、颇具研究价值。

与初唐第220窟主室东壁的持扇侍从相比，朱色大袖衫、合体绿色小袖、裲裆、袴褶的造型基本没有变化，说明当时依然延续着魏晋、隋、初唐以来皇家仪仗卫队的标准搭配。相比第220窟的侍从，这身造型又增加了肩上的饰物、丰富了服装上的装饰，但具体的结构与工艺还有待研究。绘画整理时，尽量按照目前洞窟中所见还原。

（文：李迎军）

Compared with Vimalakirti, Manjusri, Sariputra, Bodhisattva, emperor, ministers and princes of various countries, the fan-holding servant in the Vimalakirti Sutra illustration is a "little man", but this little man is an image that frequently appears beside the emperor in the Vimalakirti Sutra illustration in various caves. Although the color of the fan-holding servant on the east wall of the main chamber in Cave 103 at Mogao Grottoes is simple, while the shape is lively, which has great research value.

Compared with the fan-holding servant on the east wall of the main chamber in Cave 220 dated to the early Tang Dynasty, the shapes of vermilion large-sleeved Shan, green small-sleeved shirt, Liangdang and pleated pants basically remained unchanged, indicating that the standard arrangement of royal guard of honor since the Wei, Jin, Sui and early Tang Dynasties was still used at this time. At the mean time this servant has been added ornaments on the shoulders and enriched the decoration of clothes, but the specific structure and technique need to be studied. When doing the illustration, the painter tried to restore the details according to current painting in the cave.

(Written by: Li Yingjun)

图: 李迎军 Painted by: Li Yingjun

The prince's clothes
on the south side of the east wall
in the main chamber of Cave 103
dated to the high Tang Dynasty
at Mogao Grottoes

王子服饰
莫高窟盛唐第103
窟主室东壁南侧

从各国王子的人数与绘画的精细程度看，莫高窟第103窟主室东壁绘制的这幅壁画在诸多维摩诘经变画中并不突出，但其中有几位王子形象相对独特，画面右端穿着翻领红袍的王子就是其中之一，他的发型、服装服饰与陕西乾陵章怀太子墓墓道壁画《礼宾图》中的使者形象接近，多认为是罗马使者。

这位王子深目高鼻，留着齐耳的卷发，无髻无冠，身上穿着窄袖对襟翻领长袍，袖口、衣襟有边饰，侧缝下端有开衩。由于对襟翻领袍内穿有半臂，所以宽松袍服的肩部被里面挺括的半臂撑起，形成独特的廓型，近似于现代西方服装中加了垫肩的服装造型。半臂在唐前期曾广泛流行，多套穿在上襦的外面，有大量的传世绘画、雕塑都如实反映了这一穿着方式。相比而言，把半臂穿在长袍里面的造型数量较少，大多出现在西域胡人俑上。

在"维摩诘示疾"下方的各国王子中，第二排里侧的王子头戴卷檐毡帽，身穿翻领对襟袍，袍长至膝下，足穿长筒靴，前侧靴面有独特的毛皮装饰，肩上搭着粮食袋。这位王子由于位列后排，而没有被详细刻画，他站立的姿势、服装的形态也被前排的王子大面积遮挡，但在仅见的寥寥数笔线条中，帽子、衣襟、袖口等结构均得以清晰表达，靴子上毛皮装饰与肩头粮食袋的细节更是概括得简练精到。

由于当时的画师只用线条表达了服装结构，并没有为这一被遮挡的造型着色，所以无法判断王子所着翻领对襟袍的服色，绘画整理时如实地遵循了壁画上的色彩关系，不强调长袍的服色，而是以突出粮食袋、饰毛皮的靴子为主旨。

（文：李迎军）

Judging from the number and the fineness of princes, this mural on the east wall of the main chamber in Cave 103 at Mogao Grottoes is not prominent in many other Vimalakirti Sutra illustration paintings, but several princes here have relatively unique appearances. The prince in red lapel robe at the right end of the painting is one of them. His hairstyle, clothes and decorations are close to the envoy image in *Li Bin Tu* from Prince Zhang Huai tomb in Qianling, Shanxi Province, they are mostly considered to be Roman envoys.

The prince has deep eyes and a high nose, ear length curly hair, no hair bun and no crown, wearing narrow-sleeved lapel robe on the upper body. The cuffs and trims are decorated with patterns, and there are slits at the lower end of the robe. Because half-sleeved coat is worn in the lapel robe, the shoulders of loose robe are supported by the hard half-sleeved coat inside to form a unique shape, the effect is similar to modern western clothes which have shoulder pads. Half-sleeved coat was widely popular in the early Tang Dynasty, they were usually worn outside the upper Ru shirt. A large number of existing ancient paintings and sculptures truthfully reflected this way of wearing. In contrast, it's not common to wear half-sleeved coat under the robe, most of which appear on the foreign figurines from the Western Regions.

Among the princes of various countries below the "Vimalakirti pretending sick", the prince in the inner side of the second row wears a rolled-brim felt hat and lapel robe. The robe beyond the knees, and feet in long boots, which have unique fur decoration on the surface and a grain bag on his shoulders. The prince was not depicted in details because he stands in the back row. His standing posture and the shape of his clothes are also mostly covered by the princes in the front row. However, by few lines, the structures such as hat, hem and cuffs are clearly expressed, and the details of fur decoration on boots and grain bag around neck are concise and clear.

Because the painter just depicted the clothing structure by lines and did not color this blocked figure, we do not know the color of the lapel robe worn by the prince. When doing the illustration, the painter truthfully followed the color on the mural, who did not emphasize the color of the robe, but focused on the food bag and fur boots.

(Written by: Li Yingjun)

图：李迎军　Painted by: Li Yíngjun

The believers' clothes
on the south wall of the main chamber
in Cave 103 dated to the high Tang
Dynasty at Mogao Grottoes

信徒服饰

莫高窟盛唐第103窟主室南壁

此图表现盛唐第103窟南壁法华经中"方便品"绕塔供养的场景。"方便品"偈语中说："如果有人在佛塔前礼拜，或者合掌、举手、低头，都是功德，都可以成佛。"画师在山前旷野中画了一座白塔，塔前八名信徒正在进行礼拜，或合掌，或行五体投地礼，或胡跪，复原的两位信徒一人合掌平拱，一人举手致敬，神态十分虔诚，展现了一个庄严的礼佛场景。

两位信徒均戴软质幞头，后垂两脚较短。戴幞头的人群不分贵贱，官员、士人和劳动者均可佩戴。信徒均着襕衫，腰束革带，足蹬乌皮靴，为唐代男子官员士人之常服。

（文：董昳云、吴波）

This painting shows circumambulation worship around stupa from "Upaayakausalyaparivartah" of the Saddharmapundarika Sutra illustration on the south wall of Cave 103 dated to the high Tang Dynasty. The verse in "Upaayakausalyaparivartah" says: "If someone worshiped or put hands together, or hands raised or head lowered in front of stupa, these actions are all good merits and virtues, and people who did these all can become Buddha." The painter painted a white stupa on the ground in front of mountains. The eight believers in front of the stupa are praying, some put their hands together, some kowtow, or foreign kneel. The two believers one put hands in front chest and the other raised his hand up to pay respect. They look very pious, showing a solemn scene of Buddha worshiping.

Both believers wear soft Fu hat with short ends hanging behind them. Everybody could wear Fu hat, officials, scholars and workers, regardless their social status. The believers all wear Lanshan, leather belt at waist and feet in black leather boots, which were often worn by male officials and literati in the Tang Dynasty.

(Written by: Dong Yiyun, Wu Bo)

The donors' clothes
on the east side of the south wall in the
main chamber of Cave 103 dated to the
high Tang Dynasty at Mogao Grottoes

供养人服饰
莫高窟盛唐第103窟主室南
壁东侧

此图位于第103窟主室南壁东侧，画面最上端描绘帝释天得佛授陀罗尼经加归天界，得到解脱转世之苦的善住与帝释天回访世尊，世尊摩善住顶授菩提记的场景。其下方描绘象征陀罗尼经的日轮、安置陀罗尼经的宝塔、高幢、高山及礼拜的人们。复原的一男两女供养人均双手合十，微微弯腰，神态虔诚，周围还有数位礼佛的男女供养人。

左首男供养人头着乌纱巾，也称"乌匼巾""小乌巾"，是以黑色纱罗制成的头巾，多用于文人隐士，身穿对襟浅色襦裙，袖祛、领缘和腰带为蓝色，袖祛较小，如黄牛下垂的肉皱，袖型为垂胡袖，足蹬云头履。后两位女供养人头梳惊鹄髻，整体服饰形制相近，均穿着袿衣，袖型为垂胡袖，浅襦之外披褐色半袖，半袖袖缘呈绿色伞状，未穿蔽膝，但有两根华髾飘出。

（文：董昳云、吴波）

This painting is located on the east side of the south wall in Cave 103. At the top of the painting, it depicts the scene that Śakro devānām indrah was granted the Sarvadurgatipariśodhana Uṣṇīṣa Vijaya Dhāraṇī sūtra by Buddha and returned to heaven, then celestial being prince Shanzhu ended the pain of reincarnation, He and Śakro devānām indrah together visited Buddha. Buddha touched Shanzhu's head and gave Bodhi vyakarana, and the lower part painted sun wheel symbolizing Dhāraṇī sūtra, and stupa, high buildings, high mountains where Dhāraṇī sūtra can be placed, also with worshipers. A man and two women worshipers are all put their palms together, bow slightly, look pious, around them are several male and female Buddha worshipers.

The man on the left wears Wushajin on head, also known as "Wuzajin" or "Xiaowujin". It is a headscarf made of black gauze, mostly used by literati and hermits. He wears parallel-lapel light color Ru skirt, the cuffs, collar and belt are blue, and the cuffs are small, the sleeves just like the sagging meat wrinkles of yellow cattle, which was so called drooping foreign sleeves, and his feet in cloud head shoes. The two women behind combed their hair in Startled Swan hair bun who have similar dress, both wear Guiyi with drooping foreign sleeves, covered with brown half-sleeved coat outside. The cuffs of the half-sleeved coat is in the shape of a green umbrella, without Bixi, but two streamers can be seen.

(Written by: Dong Yiyun, Wu Bo)

图：吴波　Painted by: Wu Bo

此图选取自盛唐第103窟主室东壁门上部维摩诘经变之"佛国品"。经文曰："毗耶离城有一长者子，名曰宝积，与五百长者子俱持七宝盖诣佛所，各以其宝盖供养佛。佛以威神，令诸宝盖，合成一盖，遍覆三千大千世界。"画工据此经文，画一大宝盖，盖下有说法图一铺，释迦牟尼居中说法，左下侧是各持一个小型宝盖的三位王子，供养于佛。

王子们头戴幞头，有两脚垂于脑后，现已褪色。三位王子俱身穿袴褶，上身着齐膝大袖衣，下着宽口裤。在唐代，袴褶为皇帝贵族、文武官员皆穿着的服饰，常与平巾帻搭配，《旧唐书·舆服志》中有："若服袴褶，则与平巾帻通著。"《旧唐书》规定了其服用制度："袴褶之制：五品以上，细绫及罗为之，六品以下，小绫为之，三品以上紫，五品以上绯，七品以上绿，九品以上碧。"壁画以三位王子礼佛的角度，展现了袴褶背面的形制。

（文：董昳云、吴波）

The princes' dress
on the upper part of the east gate in the main chamber of Cave 103 dated to the high Tang Dynasty at Mogao Grottoes

王子服饰

窟主室东壁门上部

莫高窟盛唐第103

This painting is selected from the "Chapter of Buddhist World" of the Vimalakirti Sutra illustration at the upper part of the east gate in the main chamber of Cave 103 dated to the high Tang Dynasty. The Scripture says: there is an elder's son named Baoji in Vaishali city. Together with the five hundred elder's sons, they hold seven-treasure canopy to Buddha's place, and everybody used their canopy to offer Buddha. Buddha used His magical power to combine all the treasure canopies into one, which can cover the Great Chiliocosm. Based on the Scripture, the painter drew a large treasure canopy, under the canopy is a Dharma assembly. Buddha Sakyamuni is in the middle, on the lower left are three princes, each holding a small treasure canopy to offer Buddha.

The princes wear Fu hat and have two ends hanging behind their heads, which are faded now. The three princes are all wearing Kuxi, knee length long-sleeved coat and big cuffs pants. In the Tang Dynasty, Kuxi were worn by emperors, nobles, civil and military officials, and often matched with Pingjinze. *The Old Tang Book · Yu Fu Zhi* said: "If you wear Kuxi, then match with Pingjinze." *The Old Tang Book* also stipulated its ranking system: "The regulation of Kuxi: above Level five, fine Ling and Luo, below Level six, small Ling, more than Level three are purple, more than Level five are crimson, more than Level seven are green and more than Level nine are bluish green." The mural shows the back details of Kuxi of the three princes who are worshiping Buddha.

(Written by: Dong Yiyun, Wu Bo)

此图选取自盛唐第103窟主室东壁门上部维摩诘经变之"佛国品"。释迦牟尼居中说法，右下侧为维摩诘。魏晋时期，《维摩诘经》广为流传，"辩才无碍"的维摩诘成为崇尚思辩的清谈学者的崇拜榜样。作为在家修行的居士，维摩诘的衣着打扮与文人士大夫有着相似之处或者即是以他们的形象体现。

此身维摩诘为东壁门上部南侧"维摩诘示疾"中维摩诘的缩小版，服饰均为内着曲领中单，白色袴褶，外披鹤氅裘。南朝宋时刘义庆《世说新语·企羡》中有："孟昶未达时，家在京口，尝见王恭乘高舆，被鹤氅裘。氅是斗篷、披风之类的御寒长外衣，鹤氅，即一块用仙鹤羽毛做的披肩。"在晚唐第9窟维摩诘经变画中，能够清晰地观察到画匠用白色的排线在氅上刻画仙鹤羽毛，表现出士大夫瘦骨清相、超然无为的服饰风尚。

（文：董昳云、吴波）

The Vimalakirti's clothes
on the upper part of the east door in Cave 103 dated to the high Tang Dynasty at Mogao Grottoes

维摩诘服饰

窟主室东壁门上部

莫高窟盛唐第103

This painting is also taken from the "Chapter of Buddhist World" of the Vimalakirti Sutra illustration on the upper part of the east door of Cave 103 dated to the high Tang Dynasty. Buddha Sakyamuni sits in the middle, and the lower right is Vimalakirti. During the Wei and Jin Dynasties, Vimalakirti Sutra was widely spread, and the "eloquent" Vimalakirti became an idol for scholars who advocated idle talk. As a upasaka, Vimalakirti's clothes are similar to those of literati and officials, or this image is their image.

This Vimalakirti image is a smaller version of Vimalakirti in the "Vimalakirti pretending sick" on the south side of the upper part of the east door. He wears curved collar Zhongdan inside, white Kuxi, covered Hechangqiu outside. In Liu Yiqing's *A New Account of the Tales of the World* in the Southern Song Dynasty, "when Meng Chang did not arrive, he lived in Jingkou, always saw Wang Gong in high vehicle and covered with Hechangqiu. Chang is a long coat like cloak or robe to keep out the coldness, so Hechang means crane cloak which is a shawl made of crane feathers." In the Vimalakirti Sutra painting in Cave 9 of the late Tang Dynasty, it can be clearly seen that the painter used white lines to depict crane feathers on the cloak, showing the fashion of thin bones, clear appearance and detached and inaction of literati.

(Written by: Dong Yiyun, Wu Bo)

图：吴波　Painted by: Wu Bo

Dunhuang Mogao Grottoes Cave

113 of the High Tang Dynasty

敦煌莫高窟盛唐

第113窟

　　第113窟为盛唐时期开凿的覆斗顶洞窟，经五代重修。主室窟顶藻井为团花井心，四披画千佛。西壁开盝顶帐形龛，彩塑一佛、二弟子、二菩萨。盝形龛顶为棋格团花，四披上画药师立佛共二十一身，下画团花。龛内四壁画屏风画共六扇，屏风内画佛弟子。龛沿画卷草和半团花图案。龛外南北侧台上各塑一天王。南壁画观无量寿经变一铺，中间画净土庄严像，东侧画两列十六观，西侧画一列未生怨。北壁画弥勒经变一铺。两壁西侧各画夜叉一身。东壁门上和门北画法华经变观音普门品，门南画千手千眼观世音一铺。前室及甬道存五代壁画，西壁门上画五代重修愿文榜题，南壁五代画一佛，正中开龛，甬道五代画药师经变。

　　此窟壁画大约绘于盛唐晚期，即吐蕃占领敦煌前，此时已经没有盛唐前期的鲜艳富丽，但窟内画面布局及题材已经非常成熟了。壁画题材分别为观无量寿经变、弥勒经变及观音普门品，反映出当时流行的信仰。窟内的彩塑和壁画布局为典型的盛唐特征。彩塑保存较好，人物造型丰满健硕、姿态优美、表情生动，极具感染力，体现出唐代饱满而富有张力的审美风格。彩塑的服饰上有逼真的染织图案，纹样精美，手法写实，可见当时的染织工艺已达到很高的水平，凸显出盛唐艺术灿烂精致的风格。窟内人物形象精美，服饰结构清楚、图案丰富、色彩沉稳，是研究盛唐服饰的重要资料。

（文：杨婧嫱）

Cave 113 is a truncated pyramidal ceiling cave excavated in the high Tang Dynasty, which has been repaired in the Five Dynasty. The caisson at the top of the main chamber has round flower pattern as the center, with four slopes painted with Thousand Buddhas. On the west wall has a tent shaped niche, with painted sculptures of one Buddha, two disciples and two Bodhisattvas. On the flat checks niche ceiling filled with round flower pattern, and the four slopes of the niche are covered with twenty-one standing Bhaisajyaguru Budhha, below them are painted round flower pattern. The four walls of the niche have six pieces of folding screens painting, and Buddhist disciples painted inside. Niche edges are painted with scrolling vine pattern and half round flower pattern. Outside the niche, one Maharāja-deva painted sculpture is each on the north and south side platforms. The south wall painted Aparimitayur-sutra Sutra illustration, with the Western Pure Land in the middle, Sixteen Visualization on the east side and Ajatasattu's story on the west side. The north wall is painted with Maitreya Sutra illustration. The west side of the two walls each painted a Yakṣa. Above and north of the east door painted the Saddharmapundarika Sutra, the chapter of Samantamukhaparivartah, and on the south side of the door is painted with thousand hands and thousand eyes Avalokitesvara. There are the Five Dynasty murals in the antechamber and corridor, above the door on west wall in the antechamber has inscription about restoration dated to the Five Dynasty, and a Buddha is painted on the south wall dated to the Five Dynasty, also has a niche in the middle. In the corridor is painted with Bhaiṣajyaguruvaiḍūryaprabhāsapūrvapraṇidhānaviśeṣavistara sūtra illustration dated to the Five Dynasty.

The murals in this cave were painted in the later time of the high Tang Dynasty, that is, before Tubo occupied Dunhuang. At this period, there was no brightness and richness feeling of the early time of the high Tang Dynasty, but the painting layout and theme in the cave are very mature. The themes of the murals are Amitayurdhyana Sutra, Maitreya Sutra and the chapter of Samantamukhaparivartah, reflecting the popular beliefs at that time. The layout of painted sculptures and murals in the cave is a typical feature of the high Tang Dynasty. The painted sculptures are well preserved, the figures are plump and vigorous, beautiful postures, vivid expressions and very infectious, reflecting the full and tension aesthetic style of the high Tang Dynasty. There are clear dyeing and weaving patterns on the clothes of painted sculptures, with exquisite patterns and professional techniques. It can be seen that the dyeing and weaving technology at that time had reached a high level, showing the brilliant and exquisite style of the art in the high Tang Dynasty. The figures in the cave are exquisite, the clothing structures are clear, the patterns are rich and the colors are calm, these are important materials for the study of the high Tang Dynasty clothes.

(Written by: Yang Jingqiang)

The Bodhisattva's clothes
on the south wall in the main chamber of
Cave 113 dated to the high Tang Dynasty
at Mogao Grottoes

菩萨服饰
莫高窟盛唐第113窟主室南壁

盛唐第113窟主室南壁菩萨，面容温婉，眉清目秀，高鼻小口，双手合十，交腿端坐于莲花台之上。上身天衣绕身，下身穿黄色长裙，四瓣花纹点缀其间，腰部装饰以腰襻，并有裙带垂落。菩萨头戴莲花宝冠，冠体正中莲花上镶嵌着绿宝石，两侧装饰有莲花宝珠，并有冠缯与垂坠从头后两侧飘拂，与珠宝项链和手镯相映成趣。

在画面的艺术处理上，菩萨的静态造型与其天衣、飘带的动态形成鲜明的对比，给人以静中有动、动静结合的视觉效果。同时，依据不同质地的织物选择不同的用线方式，天衣、裙子和飘带的用线均有微妙的区别，使织物的肌理与线条之间协调统一，追求完美的艺术表达力与感染力。

（文：刘元风）

The Bodhisattva on the south wall in the main chamber of Cave 113 dated to the high Tang Dynasty has a gentle face, beautiful eyebrows, a high nose and a small mouth, His hands folded and legs crossed sitting on the lotus seat. The upper body is wrapped in silk shawl, and the lower body is dressed in a long yellow skirt, dotted with four-petal flower pattern. The waist is decorated with silk belt and falls naturally. The Bodhisattva wears a lotus crown with emeralds inlaid on the lotus in the middle of the crown body. Lotus jewels are decorated on both sides, and there are crown laces and pendants floating on both sides from the back of the head, which are corresponding interestingly to necklaces and bracelets.

In the artistic processing of the picture, the static shape of the Bodhisattva is in sharp contrast to the dynamics of His silk shawl and ribbons, giving people a visual effect of moving in static and combining dynamic and static. At the same time, different ways of using lines are applied according to different textures of fabrics. The threads of silk shawl, skirt and ribbons are slightly different, so as to coordinate and unify the textures and lines of the fabrics and pursue perfect artistic expression and appeal.

(Written by: Liu Yuanfeng)

Painted by: Liu Yuanfeng

The Bodhisattva's clothes on the north side of the west niche in the main chamber of Cave 113 dated to the high Tang Dynasty at Mogao Grottoes

彩塑菩萨服饰 莫高窟盛唐第113窟主室西壁龛内北侧

盛唐第113窟主室西壁龛内北侧彩塑菩萨，上半身披络腋，下半身穿薄纱落地长裙，裙子的结构呈现总体而有节奏的褶皱效果，并有卷草纹样点缀其间，腰系绣花罗裙，裙子前面有波浪状边饰盘旋而下，更显其裙子的律动美感。菩萨头束球形髻，佩戴宝珠装饰的臂钏和手镯，跣足立于莲花台上。

在彩塑的艺术处理上，与当时的壁画人物相一致，其形象丰满圆润，姿态优雅，整体造型风格虽然受到印度佛教造像的影响，但更多地追求在深厚的文化底蕴的基础上去塑造具有中国古典韵味的审美意境，使我们看到莫高窟盛唐雕塑艺术在审美视域的进一步拓展的趋势。

（文：刘元风）

The painted sculpture Bodhisattva on the north side of the west niche in the main chamber of Cave 113 dated to the high Tang Dynasty, His upper body covered by Luoye and the lower body wearing a tulle long floor skirt. The design of the skirt presents an overall and rhythmic wrinkle effect, and dotted with scrolling vine pattern. The waist is wrapped by an Luo skirt embroidered flower pattern, and there are wavy trims spiraling down in front of the skirt, which show the rhythmic beauty of the skirt. The Bodhisattva has a ball hair bun, armlets and bracelets decorated with precious jewels, and stands barefoot on the lotus platform.

In terms of the artistic treatment of this painted sculpture, it consistent with the mural figures at that time. Its image is plump and mellow, and the posture is elegant. Although the overall modeling style is influenced by Indian Buddhist statues, while it pursues more to create an aesthetic artistic conception by Chinese classical charm on the basis of profound cultural heritage, allow us to see the trend of further expansion of the aesthetic horizon of sculpture art during the high Tang Dynasty at Mogao Grottoes.

(Written by: Liu Yuanfeng)

The Maharāja-deva's clothes
on the north outside the west niche in the main
chamber of Cave 113 dated to the high Tang
Dynasty at Mogao Grottoes

天王服饰
莫高窟盛唐第113窟主室西
壁龛外北侧

这身高大勇猛、威严孔武的天王塑像立于西壁龛外北侧，在类似的天王像中，天王的双手多挂长剑，该像双手搭叠在腹前也呈挂剑状，因此推断原塑像中应有长剑。这身天王须发皆赤，穿戴护项、披膊、兽头护肩、护臂、明光甲、束甲带、护腹、腿裙、碎花白袴、皮靴。唐时期彩塑的写实手法日趋精湛，塑像的造型特征、精神风貌都更加生动传神，服装更是刻画得细腻自然。以这身着甲胄的天王像为例，立体的塑像准确地呈现了多层甲衣的结构关系，较壁画中的天王像更清晰翔实。

在这身天王的立体甲衣造型中，并没有甲叶结构的塑造，取而代之的是饱满的图案与艳丽的施彩，金色的边线也起到画龙点睛的装饰作用，可见这身天王穿着的甲衣并不具备实战功能。在《唐六典》记载的13种甲衣中，白布甲、皂绢甲、布背甲都是用布料制作，以布制甲兼具装饰性强、轻巧舒适等特点，主要用于礼仪性场合。

（文：李迎军）

The tall, brave and majestic painted sculpture Maharāja-deva is standing on the north outside the west niche. In similar sculptures of Maharāja-deva, their hands usually rest on a long sword. The sculpture's hands are overlaid in front of the belly seem like resting on something. Therefore, it is inferred that there should be a long sword in the original sculpture. The Maharāja-deva has red beard and hair, wears neck armor, arm covers, beast head shoulder covers, arm armor, Mingguang armor, armor fasten belt, belly armor, battle skirt, broken flower pattern white pants and leather boots. The realistic techniques of painted sculptures in the Tang Dynasty became more and more exquisite, the modeling characteristics and spiritual style of the sculptures were more vivid, and the clothes depiction were more delicate and natural. Taking the sculpture of the Maharāja-deva in armor as an example, the three-dimensional sculpture accurately presents the structural relationship of multi-layer armor, which is more clear and detailed than the sculpture of Maharāja-devas in wall paintings.

In the shape of the three-dimensional armor of the Maharāja-deva, there is no depict of the armor pieces structure. Instead, full patterns and gorgeous colors were used. The golden sideline is eye catching, so that the armor worn by the Maharāja-deva did not have the actual protective function. Among the thirteen kinds of armor recorded in the *Tang Liu Dian*, white cloth armor, black silk armor and back cloth armor were all made of cloth. Cloth armor has decorative, light and comfortable advantages mainly used in ceremonial occasions.

(Written by: Li Yingjun)

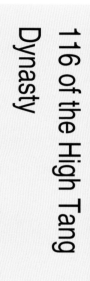

Dunhuang Mogao Grottoes Cave

116 of the High Tang

Dynasty

敦煌莫高窟盛唐

第116窟

　　第116窟为盛唐时期开凿的覆斗顶洞窟，经中唐、宋、清重修。主室窟顶藻井为中唐绘茶花环枝团花井心，四披中唐绘千佛。西壁开斜顶敞口龛，盛唐彩塑一佛、二弟子、二菩萨，清塑二菩萨。龛顶为中唐绘宝盖及飞天，宋代重描涂色。龛内西壁盛唐浮塑佛光，佛光两侧及龛内南北壁中唐绘弟子共八身，经宋代晕染。龛沿中唐绘半团花及卷草边饰，卷草边饰上色还未完成。南壁盛唐绘观无量寿经变，北壁盛唐绘弥勒经变。东壁门上盛唐绘千佛，门南、北盛唐分别画一观世音菩萨及地藏菩萨。甬道南、北壁宋各绘二身供养菩萨。

　　此窟南、北壁壁画皆由盛唐起稿，经中唐完成，宋代重描。窟内壁画线条流畅，装饰风格淡雅精致。中唐的画法比起盛唐有所变化，画师填色时为了避免颜色遮盖线条，会刻意留白，有意不将颜色填满。因此，窟内壁画的色彩也由盛唐的丰富绚丽趋向于清淡，从色彩的渐变层次仍然可以看出采用了退晕法进行绘画，人物的肌肤不以重彩，仅通过淡彩晕染仍显立体。东壁地藏菩萨及观世音菩萨，皆配璎珞，服饰配色为黑赭、石青，配色简单大方，衬托出人物优雅的气韵。

（文：杨婧嫱）

Cave 116 is a truncated pyramidal ceiling cave excavated in the high Tang Dynasty, which was repaired in the middle Tang Dynasty, the Song Dynasty and the Qing Dynasty. The caisson on the top of the main chamber has camellia ring round flower pattern in the center painted in the middle Tang Dynasty, the four slopes painted Thousand Buddhas in the middle Tang Dynasty. The west wall has a wide open niche with sloping ceiling, and in the niche there are painted sculptures of one Buddha, two disciples and two Bodhisattvas dated to the high Tang Dynasty, and two Bodhisattvas dated to the Qing Dynasty. The top of the niche was painted treasure canopy and Flying Apsaras in the middle Tang Dynasty, which were redrawn and colored in the Song Dynasty. The west niche wall has Buddha light relief dated to the high Tang Dynasty, on both sides of the Buddha light and on the north and south walls of the niche there are eight painted disciples dated to the middle Tang Dynasty, which were re-colored in the Song Dynasty. The niche edge is painted with half round flower pattern and scrolling vine pattern in the middle Tang Dynasty, and the color of scrolling vine pattern has not been completed. The south wall painted Amitayurdhyana Sutra illustration in the high Tang Dynasty, and the north wall painted Maitreya Sutra illustration in the high Tang Dynasty. Thousand Buddhas are painted above the east door in the high Tang Dynasty, and Avalokitesvara Bodhisattva and Ksitigarbharaja Bodhisattva are painted on the south and north of the door in the high Tang Dynasty. On the south and north walls of the corridor each are painted two offering Bodhisattvas.

The murals on the south and north walls of the cave were drafted in the high Tang Dynasty, completed in the middle Tang Dynasty and re-outlined in the Song Dynasty. The murals in the cave have smooth lines and elegant and exquisite decorative style. The painting method of the middle Tang Dynasty is different from that of the high Tang Dynasty. In order to avoid lines been covered during coloring, painters deliberately left some blank spaces beside lines. Therefore, the mural colors in the grottoes also tend to be lighter compare to the rich and gorgeous colors of the high Tang Dynasty. From the gradual change of color, it can still be seen that the color-gradation technique was used for painting, and the skin colors of the characters are not strong, but still have three-dimensional effect only by light color shading technique. Ksitigarbharaja Bodhisattva and Avalokitesvara Bodhisattva on east wall have keyūra, the dress colors are black, ochre and malachite green, and the color matching is simple and comfortable, setting off the elegant charm of the characters.

(Written by: Yang Jingqiang)

The Bodhisattva's clothes
on the south side of the east gate in the
main chamber of Cave 116 dated to the
high Tang Dynasty at Mogao Grottoes

菩萨服饰
莫高窟盛唐第116窟主室东
壁门南侧

盛唐第116窟主室东壁门南侧观世音菩萨，面带笑意，眼帘微垂，嘴角上扬，颈部丰满并有惯常的三道褶纹，是唐代女性以胖为美的典型象征，耳垂拉长呈矩形。

菩萨左手提净瓶，右手轻捻飘带，上身斜披络腋，络腋的正面为土红色，背面为绿色，并在胸前打结（络腋多以轻薄柔软的纱质丝织物为主，初唐时期菩萨的络腋比较宽而短；盛唐时期的络腋其装饰花纹比初唐更为丰富而华丽；中晚唐时期的络腋则趋于长而窄，装饰纹样也较少）。下身穿赭红色长裙，腰配绿色的腰裙，并有裙带打结垂落，自肩而下的长长的飘带摇曳生姿。菩萨头上佩戴日月冠、颈部、腕部佩戴珠宝首饰，"X"形长璎珞垂挂至膝部。整体服饰尽显绚丽多彩，优雅妩媚。

（文：刘元风）

On the south side of the east gate in the main chamber of Cave 116 dated to the high Tang Dynasty, Avalokitesvara Bodhisattva has a smiling face, slightly drooping eyes, mouth corners rising, plump neck with three folds. This was a typical symbol of women's beauty in the Tang Dynasty that plump is beauty, the elongated earlobes in a rectangle shape.

The Bodhisattva holds the water vase in His left hand and twiddles the ribbon by His right hand. His upper body is diagonally wrapped by Luoye. The front side of the Luoye is earth red and the back side is green, and has a knot in front of the chest (Luoye was mostly made of light, thin and soft yarn silk fabrics. Bodhisattva's Luoye of the early Tang Dynasty were relatively wide and short; decorative patterns of Luoye in the high Tang Dynasty were richer and more gorgeous than those in the early Tang Dynasty; Luoye in the middle and late Tang Dynasty tend to be long and narrow, and decorative patterns were few.) The Bodhisattva wears an ochre red skirt on the lower body and a green waist wrap on the waist, skirt band is knotted and dropped, and the long ribbons from shoulders sways. The Bodhisattva wears a sun-moon crown on the head, jewelry on the neck and wrists, and an "X" shaped long necklace hanging to the knees. The overall dress is gorgeous, colorful, elegant and charming.

(Written by: Liu Yuanfeng)

第120窟

敦煌莫高窟盛唐

120 of the High Tang
Dynasty

Dunhuang Mogao Grottoes Cave

　　第120窟是创建于盛唐时期覆斗顶洞窟，经五代、清重修。主室窟顶藻井为四瓣花井心，藻井边缘绘璎珞垂幔，四披中唐绘千佛。西壁开平顶敞口龛，清彩塑一佛、二弟子、四菩萨。龛顶中央绘说法图、两侧各绘一组赴会佛及飞天。龛内西壁盛唐浮塑佛光，佛光两侧各画二弟子，龛内南北壁各画一菩萨、二弟子。龛沿绘百花卷草边饰。东壁门上盛唐绘千佛，门南、北盛唐分别画一观世音菩萨及地藏菩萨。甬道南、北壁宋各绘二身供养菩萨。龛外南、北侧分别绘地藏一身及药师佛一身，龛下存四身供养人。南、北两壁西端各画一弟子，南壁盛唐画观无量寿经变一铺，两侧画十六观。北壁盛唐画说法图一铺。东壁门上画涅槃经变一铺，门南、门北各画一天王。前室及甬道存五代壁画。

　　此窟的涅槃经变是莫高窟盛唐时期里最小的一铺。"涅槃"为梵语音译，指的是释迦牟尼经过累世的修行，觉悟生老病死进入一种"常乐我净"的永恒境界。唐代著名高僧玄奘也将"涅槃"译为"圆寂"，即功德圆满、永恒寂静的意思。此窟涅槃经变虽小，但描绘的内容非常丰富，释迦牟尼身旁画诸菩萨、天龙八部、众弟子等的举哀图，南端画佛陀弟子先佛入灭，北端画弟子礼佛足，身后画六个弟子嚎啕哀悼。在涅槃经变中，释迦牟尼的姿势通常为右胁向下，枕右手，双足相叠横卧，佛教称为"狮子卧"。此窟涅槃经变，释迦牟尼的卧式为佛头向南，左胁而卧，与通常的形式不同。

　　此窟整体壁画风格统一，线条流畅，色彩淡雅，人物形象丰富，其中几例菩萨、天王、官员等人物形象生动，配饰鲜明，虽然颜色已氧化，但服饰结构清晰，可以作为盛唐时期人物服饰形象的参考。

（文：杨婧嫱）

Cave 120 is a truncated pyramidal ceiling cave which was built in the high Tang Dynasty and repaired in the Five Dynasty and the Qing Dynasty. The caisson at the top of the main chamber has a four-petaled flower as the center, with keyūra valance painted on the edge, and the four slopes covered by Thousand Buddhas dated to the middle Tang Dynasty. The west wall has a flat ceiling niche, with painted sculptures of one Buddha, two disciples and four Bodhisattvas made in the Qing Dynasty. The center of the niche top is painted with a Dharma assembly, and on both sides have painted a Dharma attending Buddha group and Flying Apsaras. The west wall of the niche is decorated with Buddha light relief dated to the high Tang Dynasty. Two disciples are painted on both sides of the Buddha light, and one Bodhisattva and two disciples are painted on each the north and south walls of the niche respectively. The edge of the niche is painted with multi flowers scrolling vine pattern. Thousand Buddhas are painted above the east door in the high Tang Dynasty, and Avalokitesvara Bodhisattva and Ksitigarbharaja Bodhisattva are painted on the south and north of the door in the high Tang Dynasty. On the south and north walls of the corridor are painted two offering Bodhisattvas respectively in the Song Dynasty. Outside the south and north sides of the niche are respectively painted with Ksitigarbharaja Bodhisattva and Bhaisajyaguru Buddha. Below the niche, there are four donors. One disciple is painted at the west end of the south and north walls respectively. The Aparimitayur-sutra Sutra illustration on the south wall, and Sixteen Visualization is painted on both sides. On the north wall painted a Dharma Assembly, and Nirvana Sutra illustration is painted above the east door, and a Maharāja-deva is painted on the south and north side of the door respectively. There are murals in the antechamber and corridor which painted in the Five Dynasty.

The Nirvana Sutra illustration in this cave is the smallest one in Mogao grottoes of the high Tang Dynasty. Nirvana is transliterated from Sanskrit, which refers to after numberless incarnations, Sakyamuni understood the truth of life, old age, illness and death, and into an eternal state of "eternity, happiness, self, cleanness". Xuanzang, a famous monk in the high Tang Dynasty, also translated "Nirvana" into "Complete Silence", which means complete merit and virtue and eternal silence. Although Nirvana Sutra illustration in this cave is small, the content of the picture is very rich. The mourning paintings of Bodhisattvas, Eight Divisions of Dragons and Devas and disciples are painted next to Sakyamuni Buddha. The disciple choose to die before Buddha is painted at the south end, the disciple worship Buddha's feet is painted at the north end, and six disciples mourn behind them. In Nirvana Sutra illustration, the posture of Sakyamuni is usually with His right side down, His right hand under His head, His feet overlapping and lying horizontally. Buddhists call it "Lion Lying". The Nirvana Sutra illustration in this cave is different from the usual form, Sakyamuni Buddha's head to south and left side down.

The overall mural style of this cave is unified, the lines are smooth, the colors are elegant, and the characters are diverse. Several Bodhisattvas, Maharāja-devas, officials and other characters are vivid and the accessories are clear. Although the colors have been oxidized, the clothing structure is still clear, which can be used as a reference for the characters' clothing of the high Tang Dynasty.

(Written by: Yang Jingqiang)

The Bodhisattva's clothes on the south side of the west niche in the main chamber of Cave 120 dated to the high Tang Dynasty at Mogao Grottoes

菩萨服饰 莫高窟盛唐第120窟主室西壁龛内南侧

盛唐第120窟主室西壁龛内南侧菩萨，其形象恬静而优雅，双手持长茎莲花。上半身披红蓝色络腋，下半身着花卉纹样的阔腿裤，腰间系有彩色丝带，臀部包裹蓝色腰襻，中间打结垂落两腿之间，红蓝色相间的条帛和彩带一并披挂周身飘然而下。菩萨头束云髻，发辫披肩，佩戴镶嵌绿松石的宝冠，冠缯系结分布两侧，华丽的珠宝项链和臂钏与头冠的装饰风格相一致。

在画面艺术表达上，线条自然流畅，色彩简洁淡雅，线条与色彩相互映衬。从整体布局上看，其造型、形态、色彩、线条、纹样等因素互动相融，同时丰富的造型元素也充分彰显了宗教的主题，并将其审美意识推向了"极乐世界"。

（文：刘元风）

The Bodhisattva on the south side of the west niche in the main chamber of Cave 120 dated to the high Tang Dynasty has a quiet and elegant face, holding a long stem lotus by both hands. The upper body is covered with red and blue Luoye, the lower body is covered with flower pattern wide-leg pants. His waist is tied with colored ribbons, hips are wrapped with blue silk band, the middle is knotted and hanging between the two legs, and the red and blue silk scarf and ribbons together around the body and hanging down. The Bodhisattva has a cloud hair bun, hair braids cover shoulders, and a treasure crown inlaid with turquoise and the crown laces are distributed on both sides. The gorgeous necklace and armlets are consistent with the decorative style of the crown.

In the artistic expression of the picture, the lines are natural and smooth, the colors are simple and elegant, and the lines and colors set off each other. From the overall layout, its style, shape, colors, lines, patterns and other factors interact with each other and blend. At the same time, the rich modeling elements also fully highlight the religious theme and promote its aesthetic awareness to Pure Land.

(Written by: Liu Yuanfeng)

The minister's clothes
on the east wall of the main chamber
in Cave 120 dated to the high Tang
Dynasty at Mogao Grottoes

大臣服饰

莫高窟盛唐第120窟主室东壁

1—12

在主室东壁涅槃经变中，释迦牟尼头向南、面向西，左胁而卧。佛陀身后有六位痛哭的比丘，两侧有举哀的诸菩萨、天龙八部以及世俗弟子。这身举哀的世俗弟子头戴进贤冠，身着曲领中单、对襟大袖襦、素色下裳配蔽膝，上襦下裳均有宽边缘饰，足穿笏头履，手持笏板。《新唐书·车服志》中提到："进贤冠者，文官朝参、三老五更之服也。"在唐时期的敦煌壁画中，也常见到戴进贤冠、穿上襦下裳的官员形象。因此从着装判断，此人应是举哀的朝服大臣，由于目前此身人物形象上的绝大部分颜料已经氧化变色，所以参照相关洞窟中官员的服色进行了绘画整理。

（文：李迎军）

In the Nirvana Sutra illustration on the east wall of the main chamber, Buddha Sakyamuni head towards south and face to west, lying on His left side. Behind Buddha, there are six mourning monks, and mourning Bodhisattvas, Eight Divisions of Dragons and Devas and secular disciples on both sides. This mourning secular disciple wears a Jinxian crown, a curved collar Zhongdan, a parallel-lapel large-sleeved Ru, and plain Chang on the lower body matched with Bixi. The upper Ru and lower Chang are decorated with wide trim edges, feet in Hutou shoes and holding Hu board. *Xin Tang Shu · Che Fu Zhi* mentioned: "People who wear Jinxian crown are civil servants when attending the morning court meeting with emperor, as well as Sanlao and Wugeng honorable officials." In Dunhuang murals of the high Tang Dynasty, officials wearing Jinxian crown and Ru on the upper body and Chang on the lower body are also common. Therefore, judging from the dress, this person should be a mourning minister wearing morning court meeting clothes. At present, most of the pigments on this figure have oxidized and discolored, so he is painted with reference to the dress colors of officials in relevant caves.

(Written by: Li Yingjun)

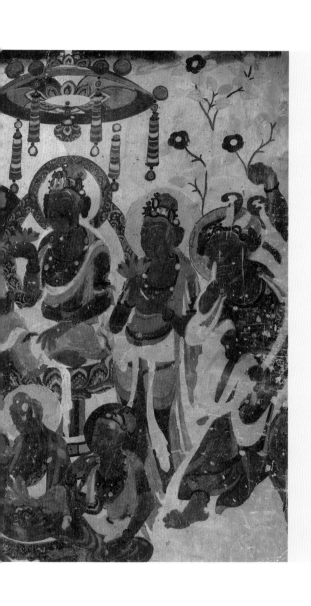

Dunhuang Mogao Grottoes Cave 121 of the High Tang Dynasty

敦煌莫高窟盛唐

第121窟

　　第121窟为盛唐时期开凿的覆斗顶洞窟，经五代、清重修。主室窟顶藻井井心已毁，藻井边缘绘璎珞垂幔，四披绘千佛。西壁开盝顶帐形龛，清彩塑一佛、二弟子、四菩萨置于马蹄形佛床上。龛顶中央五代绘棋格团花，四披五代绘药师佛及供养菩萨数身。龛内南壁、北壁五代各绘天龙八部四身、菩萨四身、夜叉一身、弟子一身。龛上五代绘山花蕉叶帐顶图案，两侧各绘两坐佛。龛下五代绘供器及供养菩萨数身。龛外南侧、北侧分别为五代绘普贤变一铺及文殊变一铺。主室南壁绘三身说法图一铺，北壁绘说法图。南、北壁底部五代补绘山水及供养人像。东壁五代绘维摩诘经变，门南绘文殊菩萨，门北绘维摩诘，门上绘佛国品，东壁底部绘供养像数身。前室及甬道存五代壁画。甬道南壁五代绘曹议金供养人像，侍从像模糊。甬道北壁绘回鹘公主等女供养人像两身。

　　此窟南壁说法图尤为精妙生动，佛陀结跏趺坐于佛座上，周围围绕着听法弟子及菩萨，众人沉浸于佛陀的教化之中，神情愉悦，甚至手舞足蹈。佛座前案上还设有供器。虽然壁画人物肤色已氧化变黑，但服饰穿搭结构及配色仍然较为清晰，仍然能看到使用了土红、石青等色彩进行装饰，为研究此时期的服饰文化提供了资料。

（文：杨婧嫱）

Cave 121 is a truncated pyramidal ceiling cave excavated in the high Tang Dynasty, which was repaired in the Five Dynasty and the Qing Dynasty. The center of the caisson at the top of the main chamber has destroyed. The edge of the caisson is painted with keyūra valance and Thousand Buddhas. A tent shaped niche was opened in the west wall, and the painted sculptures of one Buddha, two disciples and four Bodhisattvas made in the Qing Dynasty are placed on the horseshoe shaped platform. In the center of the niche ceiling painted flat checks round flower pattern in the Five Dynasty, and Bhaisajyaguru Budhha and offering Bodhisattvas are painted on the four slopes of the niche dated to the Five Dynasty. The south wall and north wall of the niche are painted with Eight Divisions of Dragons and Devas, four Bodhisattvas, one Yakṣa and one disciple dated to the Five Dynasty. Above the niche is painted with camellia banana leaves pattern in the Five Dynasty, and two sitting Buddhas are painted on both sides. Below the niche is painted with offering vessels and several bodies of offering Bodhisattvas dated to the Five Dynasty. The south and north sides outside the niche painted with Samantabhadra Bodhisattva tableau and Manjusri Bodhisattva tableau in the Five Dynasty respectively. The south wall of the main chamber is painted with three Dharma assembly in one picture, and the north wall is painted with one Dharma assembly. At the bottom of the south and north walls, landscape paintings and donor portraits were added in the Five Dynasty. The east wall is painted Vimalakirti Sutra illustration dated to the Five Dynasty, Manjusri Bodhisattva on the south side of the door, the Vimalakirti on the north side, above the door is the Chapter of Buddhist Worlds, and several donor images at the bottom of the east wall. There are murals in the antechamber and corridor dated to the Five Dynasty. The south wall of the corridor is painted with the image of Cao Yijin in the Five Dynasty, and the image of his servant is blurred. The north wall of the corridor is painted with Uighur princess two female donors.

The south wall of this cave is particularly exquisite and vivid. The Buddha sits on Buddha seat with both legs folded, surrounded by Dharma listening disciples and Bodhisattvas. People are immersed in the teaching of Buddha, look happy, and even dancing. There are offerings on the table in front of Buddha seat. Although the skin color of the characters in the mural has oxidized and turned black, their clothing structure and color matching are still relatively clear. We can still see earth red, malachite green and other colors were used for decoration, which provide information for the study of clothing culture of this period.

(Written by: Yang Jingqiang)

The Vajra Vīra's clothes
on the north wall of the main chamber
in Cave 121 dated to the high Tang
Dynasty at Mogao Grottoes

北壁

莫高窟盛唐第121窟主室

金刚力士服饰

这身愤然欲吼的金刚力士像与图中端庄持重的佛祖、菩萨形成动与静的鲜明对比。力士竖眉立目，面现怒相，左手紧握金刚杵，右手高高举起，五指张开，做震慑状，右脚着地，左脚扬起，五个脚趾张开，似乎全身的肌肉都在紧绷着用力。着装上延续了盛唐力士的典型装扮：头梳双髻，宝冠束发，冠缯飞扬，上身赤裸，腰缠锦裙，系长腰带，身上戴项圈、璎珞、手镯、脚钏、赤脚。与舞动的四肢相呼应的，是垂于腹、膝并绕臂飞扬的宽大披帛。沈从文先生在《中国古代服饰研究》中谈到披帛的形象最早出现在北朝石刻飞天像中，在现实生活中使用则是始于隋、盛于唐，并沿用后世。在敦煌壁画中，披帛在盛唐有着极具张力的艺术表现，盛唐的飞天借助漫天飘舞的披帛满壁风动，强悍壮硕的力士形象也因为融入了柔美的弧线形披帛而呈现出刚柔相济的独特韵味。

（文：李迎军）

This Vajra Vīra roars angrily whose momentum in sharp contrast with the calm and dignified Buddha and Bodhisattva in the painting. He rises eyebrows and eyes, looks angry, holds the Vajra pestle tightly in His left hand, and raises His right hand high, and stretches His five fingers, terrifying. His right foot stands on the ground, left foot raised, and the five toes stretched. It seems that His muscles of the whole body are tight and hard. The dress follows the typical Vīra dress in the high Tang Dynasty: two hair buns, hair bound by the crown, crown laces flying, the upper body is naked, the waist is wrapped in a brocade skirt, tied by a long belt, necklace, keyūra, bracelets, feet bracelets and barefoot. The wide draped silk scarf hanging on the belly, knees and flying around the arms, echoing with the dancing body. In *The Study of Ancient Chinese Clothing and Ornaments*, Mr. Shen Congwen said that the image of silk scarf first appeared in stone carving of Flying Apsaras dated to the Northern Dynasty. In real life, it began to be used in the Sui Dynasty, flourished in the high Tang Dynasty and continued to be used in future generations. In Dunhuang murals, silk scarf has a very tense artistic expression in the Tang Dynasty. The Flying Apsaras of the high Tang Dynasty fly lively with the help of silk scarf flying all over the wall. The strong and fierce Vīra also has a unique charm because of combining hardness and softness, integrated with the soft silk scarf.

(Written by: Li Yingjun)

图": 李迎军　Painted by: Lǐ Yíngjūn

The Flying Apsaras' clothes
on the north wall of the main chamber in Cave
121 dated to the high Tang Dynasty at Mogao
Grottoes

飞天服饰

莫高窟盛唐第121窟主室北壁

图中飞天位于盛唐第121窟主室北壁说法图上部，左右两侧飞天均向上飞举，呈环绕中央圣树之态。本图中所绘西侧飞天呈半坐姿态，右手持莲花，左手举花盘供养。头、颈及手臂和腕部均有璎珞装饰，上身赤裸，下着长裙，臂搭飘带，迎风飞舞。需要说明的是，该身飞天的裙摆部分的表现，洞窟壁面有褪色或颜色剥落的可能性，导致服饰结构不是十分清晰。图中有翻折以及随人体动态拉伸而呈现的褶皱，但是根据边饰来看无法完全确定其具体为裙装还是裤装，两种可能都存在。但是如果依据同时期其他洞窟或者古印度佛教视觉图像遗迹来推断的话，当时的裤子形态或许还没有完全成型，尤其是在佛教洞窟中表现飞天，往往会使用裙装的形式。因此，此身飞天的下装有可能是在风和腿部动作的共同作用下，裙下摆的结构翻折得比较复杂而已。

（文：张春佳）

The Flying Apsaras in the painting is located on the upper part of the north wall in the main chamber of Cave 121 dated to the high Tang Dynasty. The Flying Apsaras on the left and right sides both are upward in a state of surrounding the central holy tree, the one on the west side in this painting is in a semi sitting posture, holding a lotus by the right hand and holding a flower tray by the left hand. Her head, neck, arms and wrists are decorated with keyūra, the upper body is naked, the lower part is wearing a long skirt, and the arms are covered with ribbons, flying in the wind. It should be noted that the skirt color of the Flying Apsaras is faded or peeled off, so the clothing structure is not very clear. We can see folds and pleats on clothes but can not be sure that it is skirt or trousers just based on the edge decoration, both ways are possible. However, if we compare with other caves in the same period or the image relics of ancient Indian Buddhism, the shape of pants at that time may not have been fully formed, especially Flying Apsaras usually wear skirt in Buddhist caves. Therefore, this Flying Apsaras may wear skirt, just have more complex folds because the wind and leg moves.

(Written by: Zhang Chunjia)

The Flying Apsaras' clothes
on the north wall of the main chamber in Cave
121 dated to the high Tang Dynasty at Mogao
Grottoes

飞天服饰

莫高窟盛唐第121窟主室北壁

此图中飞天位于第121窟主室北壁说法图上部东侧。飞天双手持花，大体也呈半坐姿态，与西侧飞天呼应。此身飞天的下装也同西侧飞天一样有裤子的可能性，而且相比之下，由于膝盖处的结构表现，使整体看起来更像裤装。但是由于飘举的风的表现和腿部动作一同作用，大面积的裙幅也可以呈现出此类状态。只是下摆处的边饰的翻折会造成一定的混淆感，当然，这也不排除画家在绘制壁画的过程中有一些夸张丰富的可能。飘带的飞舞与流云一同造就了向上的动势，与西侧飞天一同形成围合环绕之态。画家在绘制过程中赋予衣褶和飘带极富流动感的形态特征，十分契合唐前期洞窟生动的艺术氛围。

（文：张春佳）

In this painting, the Flying Apsaras is located on the east, the upper part of the north wall in the main chamber of Cave 121. She holds flowers by both hands, and generally in a semi sitting posture, echoing with the one on the west side. The lower clothes of this Flying Apsaras has the same possibility of pants as that of on the west side. In contrast, from the knees' shape, the whole clothes looks more like trousers. However, due to the effect of blowing wind and leg moves, skirt can also has this effect. Only the folding of the hem will cause a certain sense of confusion. Of course, this does not rule out the possibility of exaggeration and diversity used by the painter in the process of painting. The flying ribbons and flowing clouds create an upward momentum and form a shape of encirclement together with the Flying Apsaras on the west side. The painter made the pleats and ribbons with dynamic, smooth and flowing features, which is very consistent with the vivid artistic atmosphere of the caves in the early period of the Tang Dynasty.

(Written by: Zhang Chunjia)

图：张春佳　Painted by: Zhang Chunjia

　　第123窟为盛唐时期开凿的覆斗顶洞窟，为武周万岁三年修造，经五代、清重修。主室窟顶藻井为团花井心，藻井边缘绘垂幔，四披绘千佛。西壁开平顶敞口龛，龛内清彩塑一佛、四弟子置于马蹄形佛床上。佛床东面绘一排男供养人。龛顶绘菩提宝盖，宝盖两侧各一飞天。龛内西壁浮塑佛光两侧绘弟子二身，南、北壁各绘弟子五身。龛沿内侧绘二方连续的菱形边饰，外侧绘二方连续的一整二半团花边饰。龛下绘建窟愿文题榜，两侧绘一排女供养人。南壁绘阿弥陀经变一铺，北壁绘弥勒经变一铺，两壁底部五代绘供养人。东壁门上绘七趺坐佛，门南、门北各绘一天王。前室及甬道存五代壁画。本窟前室与第124窟前室同为五代后周广顺三年重修。

　　盛唐时期，佛弟子的形象在莫高窟中频繁出现，在许多洞窟内可以看到佛弟子身着袈裟的形象。此窟龛内西壁保存了一身清晰的佛弟子形象，弟子外披袈裟，上以黑色线条画出田相格，格子内装饰石青、石绿、褐色的横向斑纹图案。从图案的风格来看，其配色单纯，形状简单，反映出盛唐时期佛弟子服饰图案的平面化特征。

（文：杨婧嫱）

Cave 123 is a truncated pyramidal ceiling cave excavated in the high Tang Dynasty. It was built in the third year of Wu Zhou Wansui period and repaired in the Five Dynasty and the Qing Dynasty. The caisson at the top of the main chamber has round flower pattern as center, the edge of the caisson is painted with valance, and the four slopes are painted with Thousand Buddhas. There is a niche with flat ceiling in the west wall. In the niche, there are painted sculptures of one Buddha and four disciples made in the Qing Dynasty on the horseshoe shaped platform. A row of male donors are painted on the east side of the platform. The top of the niche is painted with a Bodhi canopy, with a Flying Apsaras on each side of the cover. Two disciples are painted on each side of the Buddha light relief on the west wall of the niche, and five disciples are painted on the south and north walls respectively. Along the inner side of the niche has painted two in a group repeated rhombic ornament, and on the outer side has painted two in a group repeated a whole and two half round flower pattern edge decoration. Below the niche has inscription of prayer, and a row of female donors are painted on both sides. Amitabha Sutra illustration on the south wall, Maitreya Sutra illustration on the north wall, and at the bottom of the two walls have painted donors dated to the Five Dynasty. Above the east door have seven sitting Buddha, one painted Maharāja-deva on each side of the door. The antechamber and corridor have murals dated to the Five Dynasty. The antechamber of this cave and the antechamber of Cave 124 were repaired in the third year of Guangshun in the Later Zhou regime.

During the high Tang Dynasty, the image of Buddhist disciples appeared frequently in Mogao Grottoes. In many grottoes, we can see the image of Buddhist disciples dressed in kasaya. In the west niche of this cave has a clear image of Buddhist disciple. This disciple is dressed in kasaya, which has field pattern painted by black lines, and decorated with horizontal striped pattern of malachite green, azurite and brown. From the style of the pattern, its color matching and shape are very simple, which reflects the planarization characteristics of the dress patterns of Buddhist disciples' clothes in the high Tang Dynasty.

(Written by: Yang Jingqiang)

The disciple's clothes
in the west niche of the main chamber
in Cave 123 dated to the high Tang
Dynasty at Mogao Grottoes

弟子服饰

莫高窟盛唐第123窟主室西

壁龛内

第123窟主室西壁龛内的唐代塑像已经荡然无存，但壁上较完整地保留了唐时绘制的十二弟子像。弟子像在莫高窟唐代洞窟的西壁龛内经常出现，但这些形象的主要职能是衬托龛内佛祖、菩萨等彩塑形象。第123窟西壁龛内唐时期彩塑的遗失反而凸显出壁画上弟子像独特的艺术韵味——画面均衡错落、疏密有致，人物着色严整、惟妙惟肖，十二身弟子六人一组分列佛背光两侧，从龛的西壁分别环绕至龛南北两壁，形成了一幅构图完整、描绘传神的十二弟子群像。

或正或侧交错站立的这十二身弟子均着右祖式袈裟，十二身袈裟的整体形制与穿着方式高度统一，但颜色与图案却因人而异。这身弟子绘于龛内南壁居中的位置，双手合十、身体微前屈呈虔诚礼佛状，身着山水纹田相袈裟。画师成功地通过画面的布局位置、描绘的细致程度，以及袈裟上的山水纹样将这身弟子塑造成南壁众弟子中相对突出的形象。

（文：李迎军）

The sculptures of the Tang Dynasty in the west niche of the main chamber in Cave 123 have disappeared, but the twelve disciples painted in the Tang Dynasty have been completely preserved on the niche walls. Disciple images often appear in the west niche in the Tang Dynasty Grottoes at Mogao Grottoes, but the main function of these images was to set off the painted sculptures of Buddha and Bodhisattva in the niche. The loss of painted sculptures of the Tang Dynasty in the west niche of Cave 123 give us a chance to see the unique artistic charm of disciple images in the murals — the paintings are well balanced, and the characters are well colored and lifelike, a group of six out of the twelve disciples is arranged on each side of the Buddha's backlight respectively, covering from the west niche wall to the north and south niche walls, forming a group of twelve disciples with complete composition and vivid description.

The twelve standing disciples all wear right-shoulder-bared kasaya. The overall shape and wearing way of the twelve kasayas are highly unified, but the color and pattern vary from person to person. This disciple is painted in the middle of the south wall of the niche. His hands are folded and his body is slightly bow forward, showing a devout Buddha worshiping appearance. He is dressed in a kasaya with landscape field pattern. The painter successfully shaped this disciple into a relatively prominent image among the disciples on south wall through the layout, details, and the landscape patterns on the kasaya.

(Written by: Li Yingjun)

图": 李迎军　Painted by: Li Yingjun

　　第124窟为盛唐时期开凿的覆斗顶洞窟，经五代重修。主室窟顶藻井为团花井心，藻井边缘绘垂幔，四披绘千佛。西壁开平顶敞口龛，龛内彩塑已失。龛顶中央五代绘药师佛一铺。龛壁五代绘说法图一铺。龛沿盛唐绘石青、石绿交替的二方连续菱形边饰。龛下五代绘供器及供养菩萨数身。龛外力士台下五代绘供养人数身。龛外南、北壁盛唐绘菩萨各一身。主室南壁绘三佛说法图一铺，下五代绘男供养人数身。北壁盛唐绘阿弥陀经变一铺，下五代绘女供养人数身。东壁门上五代绘说法图一铺，两侧盛唐各绘一立佛。门南、北各绘一菩萨。前室及甬道存五代壁画，前室西壁门上有五代后周广顺三年重修愿文题榜。

　　此窟非常注重服饰的细节表现。主室西壁龛外、南北两壁、东壁两侧菩萨服饰结构清晰，表现出服饰轻薄飘逸的效果。服饰色彩鲜亮，并使用对比色进行装饰。菩萨服饰的描绘也非常注意图案的装饰，这些纹样多为十字散花纹、菱格纹，分布在络腋、腰裙、披帛、下裙上，体现出盛唐服饰纹样的华丽丰富。

（文：杨婧嫱）

Cave 124 is a truncated pyramidal ceiling cave excavated in the high Tang Dynasty, which has been repaired in the Five Dynasty. The caisson at the top of the main chamber has round flower pattern as the center, the edge of the caisson is painted with valance, and the four slopes are painted with Thousand Buddhas. The west wall has a flat ceiling niche, and the painted sculptures in the niche have lost. In the center of the niche ceiling, there is a painted Bhaisajyaguru Buddha dated to the Five Dynasty. The niche wall has a Dharma assembly painted in the Five Dynasty. The niche edge is decorated with two in a group repeated diamond pattern painted in azurite and malachite green in the high Tang Dynasty. Below the niche is painted with offering vessels and several bodies of offering Bodhisattvas dated to the Five Dynasty. Outside the niche, below the Vīra platform are painted many donors dated to the Five Dynasty. The south and north walls outside the niche each is painted with one Bodhisattva in the high Tang Dynasty. The south wall of the main chamber is painted with a picture of three Buddhas giving Dharma teaching, and below is painted many male donors dated to the Five Dynasty. The north wall is painted with Amitabha Sutra illustration dated to the high Tang Dynasty, and below is painted many female donors dated to the Five Dynasty. Above the east door is painted a Dharma assembly dated to the Five Dynasty, and each side has a painted standing Buddha dated to the high Tang Dynasty. On the south and north side of the door each has a painted Bodhisattva. There are the Five Dynasty murals in the antechamber and corridor. Above the west door of the antechamber has inscription of prayer dated to the third year of Guangshun of the Later Zhou period.

This cave's painter payed great attention to the details of clothes. Outside the west niche of the main chamber, on the north and south walls and on both sides of the east wall, the Bodhisattva clothing structures are clear, showing a light and elegant clothing effect. The dresses are brightly colored and decorated with contrasting colors. The depiction of Bodhisattva clothes also payed great attention to the decoration of patterns. Most of these patterns are cross clusters patterns and rhombic patterns, which are distributed on Luoye, waist wrap, silk scarf and lower skirt, reflecting the gorgeous and rich clothes patterns of the high Tang Dynasty.

(Written by: Yang Jingqiang)

The Bodhisattva's clothes on the south outside of the west niche in the main chamber of Cave 124 dated to the high Tang Dynasty at Mogao Grottoes

菩萨服饰

莫高窟盛唐第124窟主室西壁龛外南侧

盛唐第124窟主室西壁龛外南侧菩萨，其姿态呈"S"形，身体的重心落于左脚。上半身披红蓝相间的络腋，下半身穿土红色的阔裙。腰部附以褐色的绣花腰裙，腰裙之上配有绿色的腰襻。菩萨头顶高髻，佩戴双层绿宝石日月宝冠，垂挂珠串式璎珞、项链和臂钏，整体服饰上络腋、冠缯、腰襻与璎珞等浑然一体，华丽多彩，摇曳生姿。

从画面的艺术表现形式上来看，菩萨的"S"形体态，更多地出现在西方绘画或雕塑之中，而在中国传统绘画艺术中则较为少见。因此，这种既具古典式又有东方式的造型形象的处理，进一步显示出盛唐时期中西文化艺术交融互动在菩萨造型中的集中呈现。

（文：刘元风）

The Bodhisattva on the south outside of the west niche in the main chamber of Cave 124 dated to the high Tang Dynasty is in an "S" shape, and the body weight falls on the left foot. His upper body is covered with red and blue Luoye. The lower body is covered with a broad earth red skirt, and the waist wrapped by a brown waist wrap embroidered with flower pattern, which has a green waist silk band. The Bodhisattva has a high hair bun on His head, a double-layer emerald sun-moon crown, string keyūra, necklace and armlets. On the overall, the Luoye, crown lace, silk band and keyūra are integrated into one, gorgeous and colorful.

From the perspective of artistic expression of the picture, the "S" shaped Bodhisattva appears more in western painting or sculpture, it is rare in Chinese traditional painting art. Therefore, this treatment of both classical and oriental modeling images further show the presentation of the integration and interaction between Chinese and western culture and art in Bodhisattva modeling during the high Tang Dynasty.

(Written by: Liu Yuanfeng)

　　第125窟为盛唐时期开凿的覆斗顶洞窟，经五代、清重修。主室窟顶藻井为团花井心，藻井边缘绘垂幔，四披绘千佛。西壁开平顶敞口龛，龛内清彩塑一佛。龛顶绘云纹宝盖及二身飞天。龛壁共绘二弟子、四菩萨。龛外南侧、北侧分别绘大势至菩萨一身及观世音菩萨一身。南壁绘说法图一铺，图中为一倚坐佛、二弟子、二菩萨、二天王。北壁绘说法图一铺。东壁绘千佛。前室及甬道存五代壁画。此窟前室与第123、124窟同为五代后周广顺年间重修。

　　此窟中人物肤色虽然已氧化变黑，但人物服饰结构清晰、色彩鲜亮、图案丰富。菩萨神态温婉，头戴宝冠，身披璎珞，姿态优雅，造型精美。天王神情威武，头戴宝冠，身着甲胄，持剑或宝塔，姿态雄健。弟子面相丰圆，神态恭敬，身着纯色或云水田相袈裟。整窟人物塑造通过神情、姿态、服饰等细节体现出不同的人物形象，从而贴合人物身份，体现出画师精妙的构思与画工。

（文：杨婧嫱）

Cave 125 is a truncated pyramidal ceiling cave excavated in the high Tang Dynasty, which was repaired in the Five Dynasty and the Qing Dynasty. The caisson at the top of the main chamber has round flower pattern as the center, the edge of the caisson is painted with valance, and the four slopes are painted with Thousand Buddhas. There is a niche with flat ceiling in the west wall, in which there is a painted sculpture of Buddha made in the Qing Dynasty. The top of the niche is painted with a cloud pattern canopy and two bodies Flying Apsaras. Two disciples and four Bodhisattvas are painted on the niche wall. The south and north sides outside the niche are painted with Mahasthamaprapta Bodhisattva and Avalokitesvara Bodhisattva respectively. On the south wall is painted with a Dharma assembly, showing a sitting Buddha, two disciples, two Bodhisattvas and two Maharāja-devas. On the north wall is painted a Dharma assembly, on the east wall is painted Thousand Buddhas. There are the Five Dynasty murals in the antechamber and corridor. The antechamber of this cave, together with Cave 123 and 124, were repaired during the Guangshun period of the Later Zhou regime.

Although the skin color of the characters in this cave have oxidized and turned black, the clothes of the characters have clear structure, bright colors and rich patterns. The Bodhisattvas look gentle, wear treasure crown and keyūra, with elegant posture and exquisite shape. The Maharāja-devas look mighty, wear treasure crown, armor, and hold sword or pagoda, look strong and powerful. The disciples have round face and dignified look, and dressed in solid color or cloud water field kasaya. The figures of the whole cave reflect different characters through details such as expression, posture and clothing, so as to fit the identity of the characters which reflect the painter's exquisite conception and skills.

(Written by: Yang Jingqiang)

菩萨服饰
莫高窟盛唐第125窟主室西壁龛外南侧

The Bodhisattva's clothes on the south outside of the west niche in the main chamber of Cave 125 dated to the high Tang Dynasty at Mogao Grottoes

盛唐第125窟主室西壁龛外南侧大势至菩萨，面容娇美，妩媚动人。上半身披绿色天衣，下半身穿高腰长裙，由于受到长璎珞的兜揽作用，裙子呈现出波浪形和节奏感，并露出裙里的绿色，与外裙的土红颜色形成对比效果。白色的裙带自腰部向下在膝盖处打结，长长的红蓝丝带在身体上下左右环绕飘洒，并与底部的莲花座交织在一起。菩萨头上佩戴莲花宝冠，其中间高高卷起，两侧为火焰纹饰，并有珠串垂坠装饰，与颈部的珠宝项链和珠串璎珞相映成趣。

画面大势至菩萨的姿态在求静的同时，姿态的静与服饰的动形成了静中有动的视觉美感。在整体的艺术呈现上，佛教艺术之美得以世俗化艺术表达，佛教信仰上升为审美境界，宗教之美与艺术之美互动而交融。

（文：刘元风）

The Mahasthamaprapta Bodhisattva on the south outside the west niche of the main chamber in Cave 125 dated to the high Tang Dynasty who has a delicate face, very charming. The upper body wears a green half-sleeved Ru shirt, and the lower body wears a high waist long skirt. Due to the holding effect of the long keyūra, the skirt presents a wavy shape and rhythm, also revealed the green color inside, which formed a contrast effect with the earth red color of the outer skirt. The white skirt band is knotted at the knees from the waist down, and the long red and blue ribbons flying around the body up and down, left and right, and interwoven with the lotus cushion at the bottom. The Bodhisattva wears a lotus crown on His head, which is rolled up high in the center, decorated with flame pattern on both sides, and beads strings, this complement with the necklace and beads string keyūra on His neck.

Mahasthamaprapta Bodhisattva's posture is peaceful, while the stillness of posture and the movement of clothes form a visual beauty of movement in stillness. In the overall artistic presentation, the beauty of Buddhist art can be expressed by worldly art, Buddhist belief has risen to an aesthetic realm, and the beauties of religion and art interact and blend.

(Written by: Liu Yuanfeng)

The Vaiśravaṇa's clothes
on the south wall of Cave 125 dated to the
high Tang Dynasty at Mogao Grottoes

南壁
莫高窟盛唐第125窟主室
多闻天王服饰

主室南壁说法图遵循严格的主从序列，倚坐佛居中，护法的天王位列外侧，这身位于整铺画面最西侧、手托宝塔的天王是北方天王，又名多闻天王、毗沙门天王。作为镇鬼辟邪的守护神，天王的形象大多神武威猛、气势可畏，但这身天王却体态丰腴，面相和善，气质端肃恬淡，颇具儒将之风。天王梳髻戴宝冠，穿戴护项、披膊、明光甲、束甲带、护腹，甲衣之下露出内衬袍服宽大的袖子与衣摆，脚上穿芒鞋。

天王穿着的甲衣以现实世界的甲胄为原型，但从敦煌壁画上的天王形象看，很多天王的造型又打破了现实甲胄的完整体系，将甲衣的部分部件与其他服装混搭使用，从而塑造出源自凡间却又"非凡"的天王造型。此外，佛国世界象征飞翔的披帛与有消厄神力的摩尼宝珠也出现在天王服饰中——披帛缠绕在甲衣的皮带上，披膊上饰有火光炎炎的摩尼宝珠，超越世俗的佛国天将形象在宝物的烘托下更加栩栩如生。

（文：李迎军）

The Dharma assembly on the south wall of the main chamber follow strict layout tradition, which the sitting Buddha in the middle, and the Dharma protector Maharāja-devas on the outer area. This Maharāja-deva is located at the far west end of the whole painting and holding a pagoda who is Vaiśravaṇa Maharāja-deva, also known as Duo wen Tian Wang and Pi Sha Men Tianwang. As the Dharma protector who defeats ghosts and evil spirits, most of their images are muscular and fierce, but this Maharāja-deva has a plump body, a kind face, a quiet temperament, just like a well educated general. The Maharāja-deva has a hair bun and wears a crown, with neck armor, arm covers, Mingguang armor, armor fasten belt, and belly armor. Below the armor we can see the wide sleeves and hem of the robe, and the feet in straw shoes.

The armor worn by the Maharāja-deva took the real armor as the model, but from the images of the Maharāja-devas in Dunhuang murals, many Maharāja-devas' armor were not completely same as the real armor, instead the painter mixed and matched some parts of the armor with other clothes, to make the Maharāja-devas' clothes which originated from the mortal world but "beyond mortal". In addition, silk scarf, the symbol of flying in the Buddhist world, and Cintamani with the power to eliminate misfortune also appear in the Maharāja-devas' clothes — silk scarf is wrapped around the belt, and Cintamani with burning fire are decorated on the shoulder covers, which set off this image more celestial, more lifelike.

(Written by: Li Yingjun)

The Virūḍhaka's clothes
on the south wall of Cave 125 dated
to the high Tang Dynasty at Mogao
Grottoes

南方增长天王服饰

莫高窟盛唐第125窟主室南壁

说法图最东侧的南方增长天王竖眉怒目、气势可畏，与同壁西侧体态丰腴、沉静安详的北方托塔天王形成一动一静的鲜明对比。这种动静对比同样体现在着装上，两身天王穿着的甲衣结构、色彩与图案有很强的系列感，但细节又有微妙变化。这身南方增长天王的整体造型更加干练：梳髻戴冠、冠缯飞扬，穿着的甲衣包括护项、披膊、明光甲、束甲带、腿裙，上身甲衣之内未衬袍而直接裸露手臂，下身腿裙内露出战裙的下摆、裸小腿、戴脚钏、脚穿芒鞋。

为了凸显天王的身份，这身甲衣上也融合了披帛与摩尼宝珠。披帛缠系在腰带上，火焰升腾的摩尼宝珠装饰在双肩的披膊上。相传摩尼宝是由宝物与火焰组成，可以放射万丈光芒普照众生、消灾除困，可以实现一切净妙愿望，肩饰宝珠并升腾烈焰的造型在唐时期的天王绘画、塑像中偶有出现，应是借助摩尼宝珠体现天王的威猛神力。

（文：李迎军）

Virūḍhaka in the far east side of the Dharma assembly raised his eyebrows, eyes wide open, looks fiercely, who is in sharp contrast with the plump and quiet Vaiśravaṇa on the west side of the same wall. This dynamic and static contrast is also reflected in their dresses. The structure, color and pattern of the armor worn by the two Maharāja-devas have a strong sense of series, but the details have subtle differences. This Virūḍhaka Maharāja-deva's shape is more simpler, which is: hair bun and a crown, flying crown laces，neck protection, arm covers, Mingguang armor, armor fasten belt and battle skirt. The upper body does not have inner clothes, so we can see the bared arms, the hem of the battle skirt hanging down, with bare legs, foot bracelets, and straw shoes.

In order to highlight the identity of the Maharāja-deva, this armor is also integrated with silk scarf and Cintamani. The silk scarf is wrapped around the belt, and the flaming Cintamani are decorated on the shoulder covers. It is said that Mani treasure is composed by treasures and flames, which can radiate light to shine on all living beings, eliminate disasters and difficulties, and realize all pure and good wishes. The design of Cintamani on shoulders and rising flames occasionally appears in the paintings and sculptures of the Maharāja-devas in the Tang Dynasty, it supposed to show the mighty power of the Maharāja-devas.

(Written by: Li Yingjun)

Kasyapa's clothes
on the south wall of Cave 125 dated to the high Tang Dynasty at Mogao Grottoes

迦叶服饰
莫高窟盛唐第125窟主室南壁

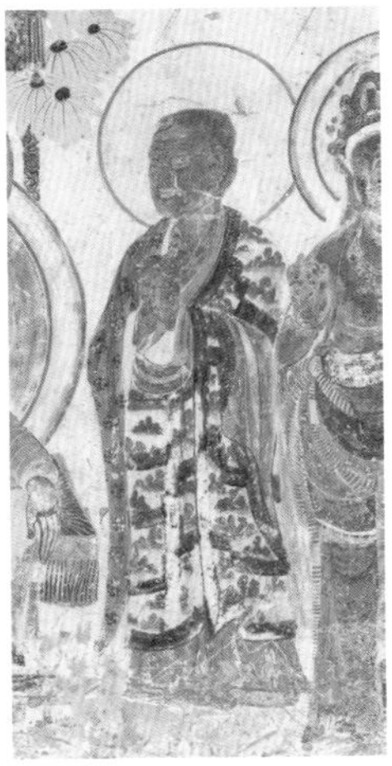

南壁说法图中的佛弟子迦叶双手握于胸前，十指交叉，似在倾听又似在发愿，身披田相山水纹袈裟，端肃虔诚地立于佛祖身侧。田相袈裟上有纵横交错的田字条相，山水纹则是袈裟上的独特纹饰——装饰有山水图案的袈裟在梁时就已经存在，是佛教汉化、世俗化的产物，这种饰有山水林木、五彩缤纷的袈裟工艺考究，价值不菲。在敦煌，山水袈裟的形象主要出现在隋、唐、宋等时期开凿的洞窟中，穿着者多为舍利弗、地藏菩萨，以及佛弟子迦叶、阿难等。

目前，壁画中这身迦叶裙下绘制的画面已经严重脱落，只能根据隐约分辨的脚的形状判断迦叶不是穿履或靴，但由于脚下还有一条深色图形（疑似鞋底造型），所以推断迦叶应不是光脚，因此，在整理时根据同洞窟中天王所穿的鞋造型绘制了芒鞋。

（文：李迎军）

In the Dharma assembly painting on the south wall, Buddhist disciple Kasyapa holds His hands in front of chest and crosses His fingers, as if listening and making wishes. He is wearing a landscape field pattern kasaya and stands piously on the side of Buddha. The kasaya has field pattern, and the landscape pattern is the unique decoration for kasaya — the kasaya decorated with landscape pattern has existed since the Liang Dynasty. It is the result of the sinicization and secularization of Buddhism. This kind of colorful kasaya decorated with mountains, rivers and trees is exquisite and valuable. In Dunhuang, the image of landscape kasaya mainly appears in the caves excavated in the Sui, Tang and Song Dynasties. Most of the wearers are Sariputra, Ksitigarbharaja Bodhisattva, Buddhist disciples Kasyapa and Ananda.

At present, the lower part of Kasyapa's skirt in the mural has disappeared. It can only be seen according to the shape of the vaguely feet that Kasyapa is not wearing shoes or boots. However, because there is a dark shape below His feet (seems like sole), so it is inferred that Kasyapa should wear something. Therefore, when doing the illustration, the straw shoes are drawn according to the Mahārāja-devas' shoes in this cave.

(Written by: Li Yingjun)

第130窟　敦煌莫高窟盛唐

130 of the High Tang Dynasty

Dunhuang Mogao Grottoes Cave

第130窟为覆斗形通顶大佛窟，东壁门上有两层明窗，盛唐时创建，经宋代重修，为盛唐时期代表窟之一，是敦煌石窟第二大窟。主室窟顶藻井为宋绘团花井心，西披绘佛光火焰，东、南、北披绘团花图案。西壁盛唐塑倚坐弥勒佛大像一尊，此大像高二十六米，被称为南大像。窟内四壁为重层壁画，南、北两壁盛唐绘高有八米的胁侍菩萨各一身，为莫高窟最大的菩萨画像。东壁表层宋绘菩萨，底层盛唐画涅槃经变。门上开上、下两层明窗，明窗内宋绘壁画。甬道顶南、北壁盛唐各开一小龛，内存盛唐及宋壁画。甬道北壁画天宝年间晋昌郡太守、墨离军使乐庭瑰及侍从供养像，南壁画乐庭瑰夫人太原王氏及其女和侍从等供养像。

此窟以甬道两壁的供养人像华贵写实而著名。南壁太原王氏都督夫人丰腴华贵，身着散花纹石榴红裙，肩披披帛，发髻簪花。身后二女身量略小，着黄裙或绿裙，披有披帛。身后侍婢均着圆领袍衫，或捧花或执扇。主尊人物头顶绘华盖，脚踏地毯，周围绘花朵，展现出礼佛的盛大场景。北壁晋昌郡太守头戴幞头，身着圆领袍衫，手持香炉。整窟壁画人物神态生动，线条流畅，与盛唐周昉人物画属同一风格，展现出盛唐世俗人物形象。

（文：杨婧嫱）

Cave 130 is a truncated pyramidal ceiling large Buddha cave with two windows above the east door. It was built in the high Tang Dynasty and repaired in the Song Dynasty, which is one of the representative caves of the high Tang Dynasty and the second largest cave at Dunhuang Grottoes. The caisson on the top of the main chamber has a round flower pattern as the center painted in the Song Dynasty, and the west slope is painted with Buddha light flame, and the east, south and north slopes painted with round flower pattern. A sculpture of Maitreya Buddha sitting against the west wall was built in the high Tang Dynasty is called the South Big Buddha, which is 26 meters high. The four walls in this cave are double-layer murals. The south and north walls each painted a 8 meters tall attendant Bodhisattva in the high Tang Dynasty, which is the largest Bodhisattva portrait in Mogao Grottoes. The upper layer of the east wall is painted with Bodhisattvas in the Song Dynasty and Nirvana Sutra illustration on the lower layer painted in the high Tang Dynasty. There are two windows above the door, in the windows have murals dated to the Song Dynasty. On the top of the south and north walls of the corridor each has a small niche excavated in the high Tang Dynasty, which preserved the high Tang Dynasty and the Song Dynasty murals. The north wall of the corridor painted the donor images of the prefect of Jinchang county and the officer of Mo Li army, Yue Tinggui and his attendants during the Tianbao period, and on the south wall painted the images of Yue Tinggui's wife, from Taiyuan Wang family, and her daughters and attendants.

This cave is famous for its magnificent and realistic portraits of the donors on both walls of the corridor. The prefect's wife on the south wall is plump and gorgeous. She is dressed in a pomegranate skirt with flower cluster patterns, wrapped in silk scarf and put flowers on hair bun. The two women behind are slightly smaller, wearing yellow and green skirt and covered with silk scarf. The servants behind them all wearing round collar robe, some holding flowers, some holding a fan. The main figure has a canopy above head, standing on a carpet which has flower pattern, showing the grand scene of worshiping Buddha. On the north wall, the prefect of Jinchang County wears Fu hat, round collar robe and holds an incense burner. The painted figures in the whole cave look vivid and lively, with smooth lines, which are in the same style as the figure paintings of Zhou Fang in the high Tang Dynasty, showing the secular figures of the high Tang Dynasty.

(Written by: Yang Jingqiang)

Prefect Yue Tinggui of
Jinchang county
offering image
on the north wall of the corridor in Cave
130 dated to the high Tang Dynasty at
Mogao Grottoes

晋昌郡太守乐庭瑰供
养像服饰
莫高窟盛唐第130窟甬道北壁

盛唐第130窟甬道北壁晋昌太守，手持长柄香炉，容貌俊朗，浓眉凤眼，胡须飘然。头戴黑色软脚幞头，身穿圆领绿色落地长袍，表现出丝质材料的垂感和飘逸感。腰间系黑色腰带，脚上着黑色软靴。其装束具有唐代典型的官府公职服饰特征。

在绘画的艺术表达上，人物的造型精确而灵动，气定神闲，画面绘制技法纯熟，色彩纯净而沉稳。特别是体现在线条的处理上，起承转合、抑扬顿挫处置得恰到好处。画风上体现出盛唐绘画艺术的南北兼容，这种绘画风格在我国绘画史上具有重要的艺术审美价值和艺术影响力。

（文：刘元风）

On the north wall of the corridor in Cave 130 dated to the high Tang Dynasty, the Jinchang county prefect holds a long-handle incense burner, who looks handsome, has thick eyebrows and phoenix eyes, long beard. He wears a black soft foot Fu hat and a round collar green long robe, which has the sense of sagging and elegance of silk materials. He also wears a black belt around his waist and black soft boots on his feet. His clothes is typical official clothes in the Tang Dynasty.

In the artistic expression of the painting, the figure shape is accurate and flexible, looks calm and free, the painting techniques are skilled, and colors are pure and calm. In particular, the lines are appropriate in the turning point, up and down. The painting style reflects the North-South painting art compatibility during the high Tang Dynasty, and this painting style has important artistic aesthetic value and artistic influence in China's painting history.

(Written by: Liu Yuanfeng)

图：刘元风　Painted by: Liu Yuanfeng

The clothes of the prefect's wife and the eldest daughter offering images on the south wall of the corridor in Cave 130 dated to the high Tang Dynasty at Mogao Grottoes

都督夫人及大女儿供养像服饰

莫高窟盛唐第130窟甬道南壁

盛唐第130窟甬道南壁都督夫人礼佛图一铺壁画，表现的是都督夫人太原王氏礼佛图中的都督夫人及大女儿的服饰。只见都督夫人王氏和大女儿身着盛装，面容圆润优美，两人均画桂叶眉，凤眼丰唇，大女儿脸上点饰有面靥，两人均束高耸的峨髻（峨髻是唐代妇女高髻的一种，因形似山峰而得名，唐李贺诗有："金翘峨髻愁暮云"的描述），髻上插饰花钗和梳篦。唐代礼制规定各级官府人员夫人的礼服颜色与官员礼服要一致，都督官位属于朝廷三品命官，因此都督夫人对礼服应采用三品官员的礼制。都督夫人手捧香炉，身穿石榴红红花碧罗曳地长裙，上穿绿色织花交领宽袖短襦，外罩绛红色底花半臂，肩披米白色披巾，腰系绿色织锦襳褵，脚穿笏头履。大女儿双手持花束，上穿红色交领宽袖短襦，下着绿色落地长裙，腰系红色襳褵，肩披白色披巾，脚穿五朵履。

整体服饰造型雍容华贵，色彩绚烂夺目，画面用线潇洒飘逸，其艺术表现风格与传世的唐人绘画名作《簪花仕女图》和《捣练图》有异曲同工之妙，在盛唐莫高窟壁画艺术中具有重要的地位。

（文：刘元风）

In the picture of prefect's wife worshiping Buddha on the south wall of the corridor in Cave 130 dated to the high Tang Dynasty shows the clothes of the prefect's wife and her eldest daughter, the wife is from Taiyuan Wang family. The prefect's wife and her eldest daughter are dressed in gorgeous clothes, with round and beautiful faces. Both of them have laurel leaf eyebrows, phoenix eyes and full lips. The eldest daughter's face is decorated with Mianye, and both of them are tied in E'ji (E'ji, Mountain peak hair bun, is a kind of high hair bun for women in the Tang Dynasty. It is named for its shape like a mountain peak. Li He's poem of the Tang Dynasty says: "the golden mountain peak hair bun envied by sunset glow"), the hair bun is decorated with hairpins and combs. The ritual system of the Tang Dynasty stipulated that official wife's dress color should be the same as that of officials. The prefect's position belonged to the third grade of imperial officials. Therefore, the prefect's wife should adopt the clothing system of third grade of officials for the dress. The prefect's wife holds an incense burner, wears a long skirt with pomegranate red and red flowers, a green woven flower cross collar wide-sleeved short Ru and a crimson color half-sleeved coat with flower pattern, a beige shawl on shoulders, green brocade Xianli(long band) on her waist, and Hu Head shoes on feet. The eldest daughter holds a bouquet by both hands, wears a red cross collar wide-sleeved short Ru, and a green trained long skirt, long Xianli around her waist, a white shawl around shoulders and feet in five petals head shoes.

The overall dress style is elegant, the colors are gorgeous and eye-catching, and the lines are natural and elegant. Its artistic expression style is similar to that of the handed down famous paintings *Group Portrait of Noble Women* and *the Picture of Making Clothes* of the Tang Dynasty. This painting plays an important role in the mural art of Mogao Grottoes of the high Tang Dynasty.

(Written by: Liu Yuanfeng)

The offering images of the prefect's wife's half-sleeved coat and silk scarf pattern on the south wall of the corridor in Cave 130 dated to the high Tang Dynasty at Mogao Grottoes

都督夫人供养像
半臂、披巾图案
莫高窟盛唐第130窟甬道南壁

都督夫人作为整幅礼佛供养图中的主体人物，其服饰最为雍容华贵。都督夫人所穿的半臂图案可谓花团锦簇，但因为覆盖在披巾下面，所以看不到整体原貌。现根据段文杰先生的壁画临摹稿及同时期图案进行变化发挥，整理为以牡丹为基础造型的团花纹。图案为四方连续式，红底上分布着盛开、半开和花苞式的红色花朵，与绿色枝叶相互映衬，显得鲜艳夺目。人物上身披着具有透明质感的米白色披巾，上面装饰着呈散点状、竖向排列的折枝花纹，每一组图案的单元纹样呈对称状，枝叶疏朗，叶片形似柳叶，其间点缀着红、黄两色的五瓣形小花，清新典雅。绘制过程中参考了常沙娜老师所著《中国敦煌历代服饰图案》一书的整理临摹稿。

（文：崔岩）

As the main figure in the whole painting of worshiping Buddha, the prefect's wife has the most elegant and gorgeous clothes. The half-sleeved coat has beautiful flower pattern, but it is covered by the the silk shawl, so we can not see the whole design. According to the mural copy of Mr. Duan Wenjie and the patterns in the same period, we design the pattern as peony based round flower pattern. The pattern is four in a group and repeated, on the red background has red flowers in full bloom, half bloom and bud shape, which are conspicuous against the green branches and leaves. The figure's upper body is covered with a beige shawl with transparent texture, which is decorated with scattered and vertically arranged small branch pattern. The pattern unit of each group is symmetrical, the branches and leaves are sparse, and the leaves are like willow leaves, interspersed with five-petaled flowers in red and yellow, which is fresh and elegant. When doing the illustration, the painters referred to the book *Clothing Patterns of China Dunhuang Mural* written by Ms. Chang Shana.

(Written by: Cui Yan)

图：崔岩　Painted by: Cui Yan

图：王可　Painted by: Wang Ke

都督夫人供养像
上襦、长裙图案
莫高窟盛唐第130窟甬道南壁

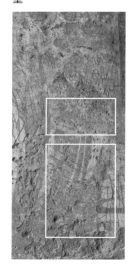

都督夫人所穿上襦和长裙的图案也以花卉纹样为主。上襦的图案为散点排列的花叶纹，花型与披巾图案相似，均为五瓣形小花，叶子分为四簇，从花朵四周呈放射状伸出。长裙图案的组织结构与披巾图案有些相似，均为呈散点状、竖向排列的折枝花纹，但叶片较为圆润，主体花型仍为五瓣形小花。两个服饰图案的主体色彩分别为青绿色和红色，配色大胆，华美艳丽但不显突兀，这是因为图案的底色和花纹色彩相互穿插呼应，青绿底上装点红色花朵，红色底上点缀青绿色的花朵和枝叶，因此取得了对比而统一的效果。绘制过程中参考了段文杰先生的壁画临摹稿和常沙娜老师的服饰图案整理临摹稿。

（文：崔岩）

The patterns of Ru and skirt worn by the prefect's wife are also mainly flower patterns. The pattern on Ru is a scattered flower and leaves cluster pattern, which is similar to the pattern on silk shawl. They are both five-petaled small flowers, and the leaves are divided into four clusters, extending radially from the flowers. The structure of the skirt pattern is somewhat similar to that of the silk shawl pattern, which are scattered and vertically arranged branch patterns, but the leaves are relatively round, and the main flower design is still five-petaled florets. The main colors of the two dress patterns are turquoise and red respectively, the color matching is bold, gorgeous but not abrupt, because the ground color and pattern color echo with each other, that means turquoise background is decorated with red flowers, and the red background is decorated with turquoise flowers, branches and leaves. Therefore, a comparative and unified effect is achieved. When doing the illustration, the painters referred to Mr. Duan Wenjie's mural copy and Ms. Chang Shana's dress pattern copies.

(Written by: Cui Yan)

图'' 王可 Painted by: Wang Ke

图'' 崔岩 Painted by: Cui Yan

The eleventh daughter's half-sleeved coat and skirt pattern on the south wall of the corridor in Cave 130 dated to the high Tang Dynasty at Mogao Grottoes

女十一娘供养像半臂、长裙图案

莫高窟盛唐第130窟甬道南壁

女十一娘供养像的服饰色彩对比鲜明、华美浓丽，与身旁着装清新淡雅的女十三娘形成了鲜明对比。女十一娘供养人像的服饰图案风格与其他几身女供养人像相似，均以散点状的花卉植物纹为主，体现了当时世俗化的审美观念。她穿着淡黄底色的半臂，上面点缀着交错排列的花叶纹，纹样为对称式的侧面视图，较为写实地画出了枝干、叶片、花蒂和花苞等整株植物的形态。色彩配置上也颇费心思，花苞部分用红色和石青色穿插变换，透露出勃勃生机。长裙图案是较为简化的散点花叶纹，纹样中心是四朵黄、红两色的五瓣形小花，四周点缀放射状叶片衬托，在石绿色的基调中表达出自然雅丽之感。绘制过程中参考了段文杰先生的壁画临摹稿和常沙娜老师的服饰图案整理临摹稿。

（文：崔岩）

The gorgeous and high contrast color of the eleventh daughter's clothes is in sharp contrast with the thirteenth daughter's dress color which is fresh and elegant. The dress pattern style of the eleventh daughter is similar to that of several other female donors portraits, which are mainly scattered flower and plant patterns, reflecting the secular aesthetic concept at that time. She wears a light yellow background half-sleeved coat, dotted with staggered flower and leaf pattern, and the pattern is side view symmetrical, which relatively painted the shape of the whole plant realistically, such as branches, leaves, flower pedicels and flower buds. The color configuration is also paid great attention. The flower bud is interspersed with red and azurite, revealing vitality. The skirt pattern is a relatively simplified scattered flower and leaf pattern, with four yellow and two red five-petaled flowers in the center, dotted with radial shaped leaves around, expressing a natural and elegant feeling in the malachite green tone. When doing the illustration, the painter referred to Mr. Duan Wenjie's mural copy and Ms. Chang Shana's clothes pattern copy.

(Written by: Cui Yan)

图：王可 Painted by: Wang Ke

The offering images of
the second daughter and
maidservant's clothes of
the prefect's wife
on the south wall of the corridor in Cave 130
dated to the high Tang Dynasty at Mogao
Grottoes

都督夫人二女儿和婢
女供养像服饰

莫高窟盛唐第130窟甬道南壁

　　盛唐第130窟甬道南壁都督夫人太原王氏礼佛图中，都督夫人的二女儿和婢女的服饰，二女儿盛装出行，面容丰满圆润，画桂叶眉（唐代妇女崇尚阔眉，桂叶眉即为阔眉的一种，眉式短而宽，因形如桂叶而得名），丹凤眼，以朱砂点唇，脸部点饰花靥（唐代妇女在盛装时，常用胭脂或丹青在脸颊、额头、眉间或太阳穴处画圆点或各种花、叶等图形，称为"花靥"），头戴凤冠，两侧各斜插步摇，同时饰花钿和角梳。身穿米白花色短襦，黄色织花长裙，外披绿色半臂，肩披蓝色披帛，腰间垂红色织锦襜褕，脚穿翘头履。

　　婢女双手托着一盘白色的茶花，头束双垂髻，身穿绿色男式圆领宽袖织花袍服，腰间有棕色系带，脚穿翘头履。

（文：刘元风）

In the picture of the prefect's wife worshiping Buddha on the south wall of the corridor in Cave 130 dated to the high Tang Dynasty, the prefect's second daughter and maidservant dress are gorgeous, with plump face and laurel leaf eyebrows (women in the Tang Dynasty advocated broad eyebrow，laurel leaf eyebrows was a kind of broad eyebrows, named for their short and wide style shaped like laurel leaves), phoenix eyes, cinnabar lips, Huaye on face (when women in the Tang Dynasty dressed up, they often painted dots or various flowers and leaves on their cheeks, foreheads, eyebrows or temples with rouge or different colors, which was called "Huaye") , a phoenix crown on head and Buyao, Huadian, and combs inserted on both sides of hair. She wears a white short Ru with flower pattern, a long yellow flower skirt, covered with green half-sleeved coat, blue silk scarf on shoulders, red brocade Xianli around waist and feet in upturned head shoes.

The maidservant holds a plate of white camellia in her hands, her hair tied into double hair buns, wearing a green men's round collar wide-sleeved flower robe, a brown belt around waist, and upturned head shoes on feet.

(Written by: Liu Yuanfeng)

The offering image of the thirteenth daughter's upper Ru and silk scarf patterns

on the south wall of the corridor in Cave 130 dated to the the high Tang Dynasty at Mogao Grottoes

女十三娘供养像上襦、披巾图案

莫高窟盛唐第130窟甬道南壁

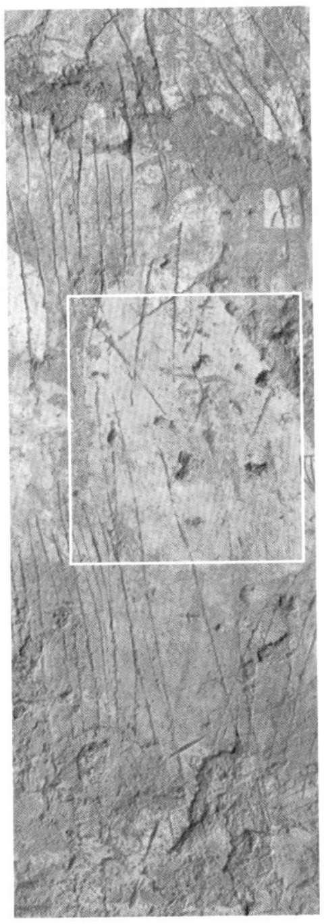

与都督夫人供养像相比，女十三娘供养像的整体服饰图案和色彩较为淡雅。她身穿米白底色的上襦，上面装饰着散点状的青绿色点花叶片纹。花纹中心为石青和石绿交错排列的三个圆点，四周伸出成组的青绿色簇状叶片。点状的叶片轻快灵动，像光芒一样围绕在圆点周围，整体形成对称的倒三角式单元造型，规整的组织结构中不乏细节变化。披巾以淡雅的石绿色为底，上面点缀着散点状的叶片纹，这种纹样与上襦中的花叶纹相似，也呈簇状伸展，小巧精致，简洁明快。绘制过程中参考了常沙娜老师所著《中国敦煌历代服饰图案》一书的整理临摹稿。

（文：崔岩）

Compared with the image of the prefect's wife, the overall dress pattern and color of the thirteenth daughter are lighter and simpler. She wears a beige upper Ru, which decorated with turquoise dotted flower leaf clusters pattern, and the center of the pattern is three roundels in azurite and malachite green stagger arranged, and groups of turquoise leaves clusters protrude around it. The dot shaped leaves are light and flexible, around the roundels like light, forming an inverted triangular symmetrical unit, and there are no lack of details in the regular structure. The silk shawl has light malachite green as background, scattered with leaf clusters pattern. This pattern is similar to the flower and leaf pattern on the upper Ru, and also extends in clusters, they are small, exquisite, simple and lively. When doing the illustration, the painter referred to the book *Clothing Patterns of China Dunhuang Mural* written by Ms. Chang Shana.

(Written by: Cui Yan)

图：王可　Painted by: Wang Ke

The offering image of
the thirteenth daughter's
skirt pattern
on the south wall of the corridor in Cave
130 dated to the high Tang Dynasty at
Mogao Grottoes

女十三娘供养像长裙
图案

莫高窟盛唐第130窟甬道南壁

女十三娘供养像身着淡黄底色的长裙，上面装饰着散点排列的花纹，清新雅丽。现参考常沙娜老师在《中国敦煌历代服饰图案》一书中的整理临摹稿，将其绘制为四方连续的折枝花纹。单元花纹为均衡结构，在倒悬卷曲的枝蔓上点缀着一朵五瓣形小花和两个小小的花苞，玲珑可爱。以折枝花纹为主体的服饰图案反复出现在都督夫人礼佛图群像中，说明这是当时女子服饰的流行纹样，也透露出盛唐时期偏爱植物和崇尚自然的审美趣味。

（文：崔岩）

The thirteenth daughter is dressed in a light yellow background long skirt, which is decorated with scattered flower pattern, fresh and elegant. We referred to the book *Clothing Patterns of China Dunhuang Mural* written by Ms. Chang Shana to draw it as four in a group repeated small branches pattern. The pattern unit has a balanced structure, with a five-petaled flower and two small flower buds on the upside down curly branches, which is exquisite and lovely. The clothing pattern with small branches as the main pattern repeatedly appeared in the group portraits of the prefect's wife worshiping Buddha, indicating that it was a popular pattern of women's clothing at that time, and also revealed the aesthetic taste of plants favored and advocating nature in the high Tang Dynasty.

(Written by: Cui Yan)

图：崔岩　Painted by: Cui Yan

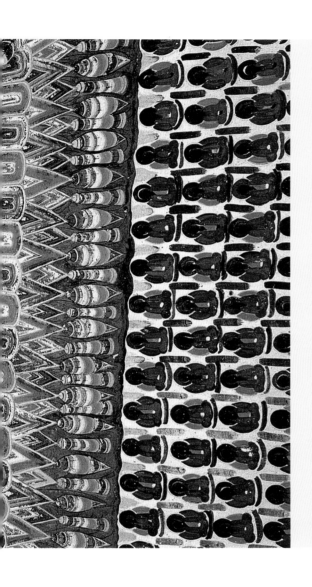

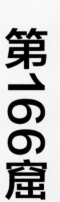

第166窟

敦煌莫高窟盛唐

166 of the High Tang
Dynasty

Dunhuang Mogao Grottoes Cave

　　第166窟为盛唐时期开凿的覆斗顶洞窟，经中唐、五代、宋、清重修。主室窟顶藻井为云头团花井心，藻井边缘绘垂幔，四披绘千佛。西壁开平顶敞口龛，龛内盛唐彩塑一佛、清修二菩萨、清塑四弟子。龛顶上绘说法图一铺，下绘菩提宝盖。龛内西壁浮塑佛光，佛光南侧画一弟子、北侧画二弟子。南、北壁各绘一菩萨、二弟子。龛沿绘菱形、团花边饰。龛下中央盛唐绘题榜，两侧宋绘供养菩萨及供养人数身。南壁西起绘一倚坐佛、七趺坐佛、七药师佛立像、三观世音菩萨，中部中唐绘二菩萨、一观世音菩萨，下宋绘一供养比丘及女供养人数身。北壁中央中唐绘说法图，下宋绘女供养人数身。东壁门南绘观世音菩萨及地藏菩萨各一身、三男供养人，宋绘菩萨及供养人。东壁门北绘地藏菩萨一身，阿弥陀佛、药师佛、多宝佛各一身，千佛数身，五代画说法图一铺，宋绘菩萨及供养人。前室及甬道存宋代壁画。此窟前室与门南下为第167窟入口，门北为第168窟入口。

　　此窟主室东壁的菩萨形象较为丰富。东壁地藏菩萨，慈眉善目，着绿色僧祇支，僧祇支边缘有画二方连续的半破团花，外披浅蓝底云水纹袈裟，其袈裟缘边用虚线装饰，模拟缝纫线迹。地藏菩萨身旁有一身观世音菩萨，观世音菩萨身姿窈窕，左手提净瓶，右手拈杨柳枝，头戴化佛冠，项饰璎珞，斜披络腋，下着长裙，轻纱透体。盛唐菩萨下装受世俗人物的影响形式多样，长裙也是其中最受欢迎的样式之一。

（文：杨婧嫱）

Cave 166 is a truncated pyramidal ceiling cave excavated in the high Tang Dynasty, which was repaired in the middle Tang Dynasty, the Five Dynasty, the Song Dynasty and the Qing Dynasty. The caisson at the top of the main chamber used cloud head round flower pattern as the center. The edge of the caisson is painted with valance and the four slopes are covered with Thousand Buddhas. There is a niche with flat ceiling in the west wall. In the niche, there are one painted sculpture Buddha dated to the high Tang Dynasty, two Bodhisattvas repaired in the Qing Dynasty and four disciples made in the Qing Dynasty. A Dharma assembly is painted on the top of the niche and a Bodhi canopy is painted below it. The west niche wall has Buddha light relief, and one painted disciple on the south side and two painted disciples on the north side. One Bodhisattva and two disciples are painted on each the south and north walls respectively. The edge of the niche is painted with diamond and round flower patterns. Below the niche, there is an inscription dated to the high Tang Dynasty in the center, and offering Bodhisattvas and some donor images on both sides dated to the Song Dynasty. On the south wall, from the west side has painted one sitting Buddha with two legs down, seven sitting Buddhas with legs folded, seven standing Bhaisajyaguru Buddhas, three Avalokitesvara Bodhisattvas. Two Bodhisattvas and one Avalokitesvara Bodhisattva are painted in the middle Tang Dynasty in the middle , and one offering Bhikkhu and female donors are painted below in the Song Dynasty. In the middle of the north wall, in the center has a Dharma assembly painted in the high Tang Dynasty, and the lower part painted some female donors in the Song Dynasty. Avalokitesvara Bodhisattva and Ksitigarbharaja Bodhisattva are painted on the south side of the east door, with three male donors, and in the Song Dynasty painted Bodhisattvas and donors. On the north side of the east door, there is a Ksitigarbharaja Bodhisattva, a Amitabha Buddha, a Bhaisajyaguru Buddha and a Prabhutaratna Buddha and Thousand Buddhas. A Dharma assembly is painted in the Five Dynasty, Bodhisattvas and donors painted in the Song Dynasty. In the antechamber and corridor have murals dated to the Song Dynasty. The entrance of Cave 167 is in the antechamber of this cave, the south side of the door, and the entrance of Cave 168 is in the antechamber of this cave, the north side of the door.

There are many Bodhisattva images on the east wall of this cave. The Ksitigarbharaja Bodhisattva image on the east wall looks kind-hearted, wearing a green Sankaksika, the edge of the Sankaksika is painted with two in a group repeated half broken round flower pattern, and covered with a light blue background cloud water pattern kasaya. The edge of the kasaya is decorated with dotted lines to simulate sewing stitches. There is a Avalokitesvara Bodhisattva beside the Ksitigarbharaja Bodhisattva, He is slim and graceful, the left hand holds a water bottle and and the right hand twiddles a willow twig, wearing Buddha image crown, keyūra, diagonally draped with Luoye, a long and transparent skirt on the lower body. Under the influence of secular clothes, the lower clothes of Bodhisattvas in the high Tang Dynasty have various forms, and the long skirt was also one of the most popular styles.

(Written by: Yang Jingqiang)

Avalokitesvara
Bodhisattva's clothes
on the south side of the east wall in the
main chamber of Cave 166 dated to the
high Tang Dynasty at Mogao Grottoes

观世音菩萨服饰
莫高窟盛唐第166窟主室东
壁南侧

盛唐第166窟主室东壁南侧观世音菩萨，面相恬静秀美，眉间点画白毫，双目微垂，朱砂点唇。左手提净瓶，右手轻拈杨柳枝。上身斜披土红和翠绿双色的络腋，下身穿土黄色轻纱透体的长裙（妇女穿裙装是从汉代以来盛饰的一种着装习俗。唐代由于社会经济、纺织技术的高度发展，文化的中外、南北交流，裙装更成为此时最具时尚风潮和最受欢迎的装束，其裙装的种类也更加丰富多彩。从敦煌壁画中可以看出，妇女的裙子长度也比以往任何时期都要长一些。同时，其腰线也进一步提高，加之唐代女性以丰满为美的倾向，提升的腰线彰显出妇女对体量感和理想化形体的审美追求），并配有绿色的腰裙，腰裙上有土红色镶边。腰部系镶嵌宝石的腰带。

观世音菩萨头戴化佛冠，冠缯与身上的飘带一起随风飘垂，身上佩戴的项链、臂钏、手镯与整体服饰相得益彰。

（文：刘元风）

The Avalokitesvara Bodhisattva on the south side of the east wall in the main chamber of Cave 166 dated to the high Tang Dynasty, has a quiet and beautiful face, Urna between eyebrows, drooping eyes and vermilion lips. His left hand lifts a water bottle and right hand pinches a willow branch. The upper body diagonally covered by earth red and emerald green Luoye, and the lower body wears earth yellow light yarn long skirt (women wear skirt had been a dress custom since the Han Dynasty. In the Tang Dynasty, due to the high development of social economy and textile technology and the cultural exchange between China and the world, north and south, skirts had become the most fashionable and popular dress at that time, and the types of skirts were more diverse. It can be seen from Dunhuang murals that women's skirts were longer than ever at that time, and the waist line also moved up. Women in the Tang Dynasty had the tendency of being plump as beauty, the raised waist line can show the sense of volume and the idealized beauty of body for aesthetic sense of women's dress), and matched with a green waist wrap with earth red trim, a belt inlaid with gemstones around the waist.

Avalokitesvara Bodhisattva wears a crown with Buddha image, crown laces and ribbons around body flying in wind. The necklace, armlets and bracelets complement with the overall dress.

(Written by: Liu Yuanfeng)

The pattern on
Avalokitesvara
Bodhisattva's Luoye
on the south side of the east wall in the
main chamber of Cave 166 dated to the
high Tang Dynasty at Mogao Grottoes

观世音菩萨络腋图案
莫高窟盛唐第166窟主室东壁
南侧

第166窟主室东壁南侧的观世音菩萨身披双色络腋，正面为土红底色，并以石绿四瓣花纹为显花，与其反面的石绿底色相呼应。花瓣用白线勾勒外围，四周及中心点缀白色圆点，呈现四方连续的排列方式。纹样色彩与底色对比强烈，主体突出，整体排列清晰，风格清新雅致。四瓣花纹样为莫高窟唐代菩萨服饰中较为常见的纹样之一，多表现于菩萨的络腋、披帛、长裙、裤或腰裙中。例如，莫高窟盛唐第66窟、第103窟、第217窟等均有此类图案。

（文：常青）

The Avalokitesvara Bodhisattva on the south side of the east wall in the main chamber of Cave 166 wears two-color Luoye, which the front side is reddish-brown, and the four-petaled malachite green flower as the pattern, that echoes with the malachite green background on the other side. The petals are outlined by white lines, and white dots are dotted around and in the center, showing a four in a group repeated arrangement. The pattern color has a strong contrast with the background color, the main body is prominent, the overall arrangement is clear, and the style is fresh and elegant. The four-petaled flower pattern was one of the common patterns on Bodhisattva clothes in the Tang Dynasty at Mogao Grottoes. It was mostly used on Bodhisattvas' Luoye, silk scarf, long skirt, trousers and waist wrap, such as Cave 66, Cave 103, Cave 217 of the high Tang Dynasty in Mogao Grottoes, etc.

(Written by: Chang Qing)

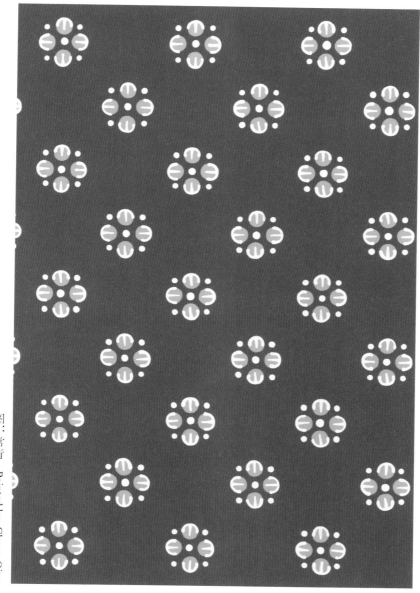

图：常青　Painted by: Chang Qing

Ksitigarbharaja
Bodhisattva's clothes
on the south side of the east wall in
Cave 166 dated to the high Tang
Dynasty at Mogao Grottoes

地藏菩萨服饰
莫高窟盛唐第166窟主室东
壁南侧

地藏菩萨是中国佛教四大菩萨之一。据佛经记载，地藏菩萨受释迦牟尼嘱托，要在释迦牟尼灭度后、弥勒佛降诞前的无佛之时留在世间济度众生。所以地藏菩萨常显示为出家僧人相，在佛教四大菩萨中只有地藏菩萨是世俗比丘形象。莫高窟第166窟主室东壁门南的观世音菩萨像与地藏菩萨像，就准确呈现了两位菩萨的这一形象特征——画面中观世音菩萨手持净瓶、着菩萨装在左，地藏菩萨圆顶光头、披右袒袈裟在右。地藏菩萨右手拇指与食指相捻结印，双指间生出云烟在两身菩萨中间升腾而上、舒展开来，云间化现七佛。

《地藏十轮经》中讲述地藏菩萨的如意宝珠可以放化光明，因此见到诸佛世界。在敦煌藏经洞帛书中，通常是通过地藏菩萨手托宝珠放光化佛来表现"地藏菩萨光中化佛"。在第166窟东壁南侧的这铺壁画中，地藏菩萨手中无宝珠，表现出的神通力是通过手指结印呈现的，这也是此铺壁画的独特之处。

（文：李迎军）

Ksitigarbharaja Bodhisattva is one of the four Bodhisattvas in Chinese Buddhism. According to Buddhist scriptures, Ksitigarbharaja Bodhisattva was entrusted by Sakyamuni to stay in the world to help all sentiment beings after Sakyamuni's death and before Maitreya's birth. Therefore, Ksitigarbharaja Bodhisattva often shows in monk appearance. Among the four Bodhisattvas in Buddhism, only Ksitigarbharaja Bodhisattva looks like a monk. The paintings of Avalokitesvara Bodhisattva and the Ksitigarbharaja Bodhisattva on the south side of the east door of the main chamber in Cave 166 at Mogao Grottoes accurately present the image characteristics of the two Bodhisattvas — in the painting, Avalokitesvara Bodhisattva holds a water bottle and wears Bodhisattva clothes on the left, and Ksitigarbharaja Bodhisattva with shaved head, wears right-shoulder-bared kasaya on the right. The thumb and index finger of Ksitigarbharaja Bodhisattva's right hand are pinched, where clouds arise from His fingers and float above the two Bodhisattvas, and seven Buddhas sit on the clouds.

In Dasacakra Ksitigarbha Sutra described that Cintamani of Ksitigarbharaja Bodhisattva can emit light, through the light people can see Buddhist worlds. In silk painting from Dunhuang library cave, it is usually through the Ksitigarbharaja Bodhisattva holding a jewel to emit light and manifest Buddhas to express the "Ksitigarbharaja Bodhisattva manifest Buddhas in light". In this mural on the south side of the east wall in Cave 166, there is no jewel in Ksitigarbharaja Bodhisattva's hands, so the magic power is presented through finger Mudra, which is also the unique part of this mural.

(Written by: Li Yingjun)

图：李迎军　Painted by: Li Yingjun

Dunhuang Mogao Grottoes Cave
171 of the High Tang
Dynasty

敦煌莫高窟盛唐

第171窟

第171窟为盛唐时期开凿的覆斗顶洞窟，经宋、清重修。主室窟顶藻井为团花井心，藻井边缘绘璎珞垂幔，四披绘千佛。西壁开盝顶帐形龛，龛内盛唐彩塑一跌坐佛、二菩萨，清塑四弟子。龛顶中央绘棋格团花及椭圆形华盖。龛内西、南、北披绘药师佛立像及供养菩萨数身，东披存药师佛立像数身。龛内西壁佛座两侧画坐佛六身，下绘莲池菩萨。南、北壁画坐佛各六身，下绘莲池伎乐。龛沿绘百花蔓草、千佛边饰。龛下宋绘供养人数身。龛外南侧绘药师佛一身。北侧绘观世音一身，下绘男、女供养人各一身。南、北、东壁各绘观无量寿经变一铺，两侧绘未生怨与十六观。前室及甬道存宋代壁画。

此窟人物虽然肤色已经氧化变黑，但是整体形象描绘得细致入微。此窟北壁阿弥陀佛身着赤色袈裟，缘边处露出里层蓝色的部分，整体衣纹线条流畅，密集而均衡。这种袈裟应为比丘三衣之一的僧伽梨，这种服饰一般按规定条数裁割再缝合，表里颜色不同。由此可见，敦煌壁画的作者绘画的素材都取自现实，他们的创作也是基于对生活的观察。

（文：杨婧嫱）

Cave 171 is a truncated pyramidal ceiling cave excavated in the high Tang Dynasty, which was repaired in the Song and the Qing Dynasty. The caisson at the top of the main chamber has the round flower pattern as the center, with keyūra valance on the edge and Thousand Buddhas on the four slopes. A tent shaped flat ceiling niche in the west wall, in which has painted sculptures of one sitting Buddha with legs folded, two Bodhisattvas dated to the high Tang Dynasty, and four disciples made in the Qing Dynasty. The center of the niche top is painted with flat checks round flower pattern and oval canopy. The west, south and north slopes of the niche have painted standing Bhaisajyaguru Buddhas and offering Bodhisattvas, and the east slope has painted several standing Bhaisajyaguru Buddhas. Six sitting Buddhas are painted on both sides of the west Buddha seat in the niche, and the lotus pond Bodhisattva is painted below. The south and north niche walls are painted with six sitting Buddhas respectively, and the lotus pond musicians are painted below. The edge of the niche is painted with flowers, vines and Thousand Buddhas patterns. Below the niche has painted some donor images dated to the Song Dynasty. The south side outside the niche is painted with Bhaisajyaguru Buddha and the north side is painted with Avalokitesvara Bodhisattva, and a male and a female donor are painted below respectively. The south, north and east walls are painted with Amitayurdhyana Sutra illustration, and Ajatasattu's story and Sixteen Visualization on both sides. There are murals dated to the Song Dynasty in the antechamber and corridor.

Although the skin color of the figures in this cave has oxidized and blackened, the images are depicted in details. Amitabha Buddha on the north wall is wearing a red kasaya，at the trim part exposed the blue inner clothes. The overall clothing lines are smooth, dense and in balance. This kind of kasaya should be Sankaksika, one of the three clothes of Bhikkhu. This kind of dress is usually cut and sewn according to the specified pieces, and the colors on the outside and inside are different. It can be seen that the painters of Dunhuang murals drew things based on reality, and their creation were also based on the observation of life.

(Written by: Yang Jingqiang)

Amitabha's clothes
on the north wall (center) of the main
chamber in Cave 171 dated to the high
Tang Dynasty at Mogao Grottoes

阿弥陀佛服饰
莫高窟盛唐第171窟主室北
壁（中央）

盛唐第171窟主室北壁阿弥陀佛，面相和蔼而安静，眉清目秀，额上点饰白毫，佛头有肉髻，留有胡须，手持花束，跏趺式端坐于莲花座之上。主尊内穿蓝色的僧祇支，外着通体的赤色袈裟，表里两层其色不一，袈裟翻卷之处显露出内里的绿色（因此，袈裟也有"复衣"之称谓）。

在绘制的艺术处理上，主尊的手的表现是一个重要问题。首先，手的结构要准确生动，用线需柔中带刚；其次，要善于深入刻画手的细部和微妙之处，如拇指和其他四个手指的相互内在关联性；再者，更为重要的是要注重手的表情达意及其与服饰的内在统一的美感与表现。

（文：刘元风）

Amitabha Buddha on the north wall of the main chamber in Cave 171 dated to the high Tang Dynasty has a kind and quiet face, clean eyes and eyebrows, Urna on the forehead. Buddha's head has Usnisa on the top, bearded, holding flowers and sitting on lotus seat. He wears blue Sankaksika inside and body-length red kasaya outside, and the colors of the outside and inside are different. The green inside of the kasaya is revealed at the turns (therefore, kasaya also has the name of "compound clothes").

In the artistic treatment of painting, the performance of Buddha's hands is an important problem. First of all, hands structure should be accurate and vivid, and the lines should be rigid within soft; second, we should depict the details and subtleties of the hand carefully, such as the internal correlation between the thumb and the other four fingers; moreover, it is more important pay attention to the meaning expressed by hands and the internal unity with the dress.

(Written by: Liu Yuanfeng)

The Bhaisajyaguru
Buddha's clothes
on the south outside the west niche in Cave
171 dated to the high Tang Dynasty at
Mogao Grottoes

药师佛服饰
莫高窟盛唐第171窟主室西
壁龛外南侧

此铺壁画位于盛唐第171窟主室西壁龛外南侧，为一尊立姿药师佛，与北侧观世音菩萨相对而立。药师佛在佛教中又称药师如来、药师琉璃光如来等，为东方净琉璃世界的教主。据《佛说药师如来本愿经》记载，念诵药师如来本愿赞，便可圆其愿，求得长寿、富贵、功名利禄等，因而在敦煌壁画中多有所现，体现出药师佛信仰的盛行。

本尊药师佛神态安详，立于莲花座上，内穿绿色僧祇支，有一条细带于腰部前中结系，外披田相纹红袈裟，袒右臂，右手自然下垂。左臂搭袈裟，左手托球形药钵，内盛救济世人的神药。整体造像神情庄静，法相慈和。

（文：董咪云、吴波）

This image is located on the south outside the west niche of Cave 171 dated to the high Tang Dynasty, which is a standing Bhaisajyaguru Buddha opposite to the Avalokitesvara Bodhisattva in the north. Bhaisajyaguru Buddha is also called medicine Tathagata and medicine glass light Tathagata in Buddhism. He is the lord of the Eastern Pure Glass World. According to Bhaiṣajyaguruvaiḍūryaprabhāsapūrvapraṇidhānaviśeṣavistara sūtra, reading and reciting the praise of Bhaisajyaguru Buddha can fulfill wishes and obtain longevity, wealth, power and fame. Therefore, Bhaisajyaguru Buddha images appear in many Dunhuang murals, reflecting the prevalence of the Bhaisajyaguru Buddha belief.

This Bhaisajyaguru Buddha looks serene and stands on a lotus seat, wearing a green Sankaksika inside, has a thin belt tied in the front and middle of the waist, and covered with a red kasaya with field pattern outside. His right arm is bare, and the right hand droops naturally. The left arm is covered by kasaya and the left hand holds a ball shaped medicine bowl, which contains the divine medicine for the relief of the world. The whole painting is caring, quiet, solemn and merciful.

(Written by: Dong Yiyun, Wu Bo)

图：吴波　Painted by: Wu Bo

The Avalokitesvara
Bodhisattva's Luoye pattern
on the north outside the west niche of the
main chamber in Cave 171 dated to the high
Tang Dynasty at Mogao Grottoes

观世音菩萨络腋图案
莫高窟盛唐第171窟主室西
壁龛外北侧

第171窟主室西壁龛外北侧的观世音菩萨，虽然肤色已经氧化变色，但整体形象的描绘精彩生动，服饰华美。此观世音菩萨面容端庄娇美，头戴化佛宝冠，左手持净瓶，身体扭动呈"S"形站姿，重心落于右脚，立于莲花之上。上身披朱红色络腋，色彩依旧鲜艳，纹样精美，在肩部打花结，并从左手腕缠绕下垂。下半身着轻薄朱色长裙，裙身虽已氧化变色，但仍可辨蓝绿花瓣相间的散花纹点缀其中。长裙外裹腰裙，腰间系绿色腰襻，外佩华丽彩绦，项链、臂钏、手镯、璎珞装点全身。现将络腋图案进行整理绘制。土黄底色上错落有致地分布着蓝色十瓣小花，花心为白色五瓣圆形小花，色彩对比明确，花型俏丽雅致，整体交错呈四方连续的排列方式。

（文：常青）

The Avalokitesvara Bodhisattva is located on the north side outside the west niche of the main chamber in Cave 171, although the skin color has oxidized and discolored, the overall image is wonderful and vivid, and the clothes are gorgeous. This Avalokitesvara Bodhisattva has a dignified and beautiful face, wears a Buddha image crown, holds a water bottle in His left hand, twists His body and stands in an "S" shape, the body weight rest on His right foot and stands on a lotus. The upper body is covered with minium color Luoye, the color is still bright, the pattern is exquisite, and tied a flower knot on the shoulder, and the Luoye is wrapped and drooped from the left wrist. The lower part of the body is wearing a light and thin vermilion skirt, although the skirt color has oxidized and changed, we can still see the blue and green petal clusters pattern. The long skirt is wrapped by waist wrap, and tied with green waist band, as well as gorgeous color rope. The whole body is decorated with necklace, armlets, bracelets and keyūra. Now we arrange and draw the pattern on Luoye. On the earth yellow background, blue ten-petal florets are stagger distributed, and the flower center is white five-petal roundels. The color contrast is clear, the flower type is beautiful and elegant, and the whole pattern is four in a group repeated stagger arranged .

(Written by: Chang Qing)

图：常青　Painted by: Chang Qing

服装复原：楚艳、崔岩

设计助理：常青、杨婧嫱、蓝津津

文字说明：崔岩、楚艳

摄影：杜帅、陈大公

化妆造型：杨树云、蓝野、周鹏、王卫艳、林颖、吴琼、张明星

模特：胡启航、王青年、陈偲昊、蓝津津、王艺璇、马祯艺、宋威葳、张翼鸥、朱震宇

Costume Reproduction: Chu Yan, Cui Yan

Design Assistant: Chang Qing, Yang Jingqiang, Lan Jinjin

Text: Cui Yan, Chu Yan

Photo: Du Shuai, Chen Dagong

Make-up: Yang Shuyun, Lan Ye, Zhou Peng, Wang Weiyan, Lin Ying, Wu Qiong, Zhang Mingxing

Model: Hu Qihang, Wang Qingnian, Chen Caihao, Lan Jinjin, Wang Yixuan, Ma Zhenyi, Song Weiwei, Zhang Yiou, Zhu Zhenyu

The Artistic
Reappearance of
Dunhuang Costume

敦煌服饰
艺术再现

The artistic reappearance of
Venerable Ananda's clothes
in Cave 45 of the high Tang Dynasty at
Dunhuang Mogao Grottoes

阿难尊者服饰艺术
再现

敦煌莫高窟盛唐第45窟

阿难尊者身着常规的比丘服饰，威仪端庄。上身内穿交领右衽半袖偏衫，领、袖缘边分别装饰着卷草纹和半团花纹二方连续图案，以印花加刺绣的工艺表现，体现了盛唐时期繁缛华丽的审美特征。偏衫大身是用印花工艺点缀的五瓣小花纹，清新典雅。下穿石绿色百褶裙，底摆处有平行的金色条纹边饰，另有印花加刺绣工艺的半团花二方连续图案装饰，与偏衫袖缘图案相呼应。外披热烈的土红色袈裟，边缘有绿色贴边，整套服装既宽绰又不乏雅致，衬托出年轻的阿难聪敏颖慧、信心满满的精神状态。

Venerable Ananda wears regular bhikkhu clothes, looks dignified. The upper body is wearing a cross collar right lapel half-sleeved shirt inside, and the collar and cuffs are decorated with scrolling vine pattern and semi-round flower pattern two in a group repeated respectively, which are expressed by printing and embroidery techniques, reflecting the complex and gorgeous aesthetic taste of the high Tang Dynasty. The main body of the shirt is decorated with five-petal small flower pattern by printing technique, which is fresh and elegant. The lower body wears a malachite green pleated skirt, with parallel gold stripes trim at the bottom, and semi-round flower pattern two in a group repeated by printing and embroidery techniques, echoing with the shirt cuffs decoration. The outside covered with earth red kasaya with green trim on the edge, and the whole set of clothes are loose and elegant, setting off the young Ananda's mental state of intelligence and confidence.

The artistic reappearance
of Prefect Yue Tinggui of
Jinchang county worshiping
group clothes
in Cave 130 of the high Tang Dynasty at
Dunhuang Mogao Grottoes

晋昌郡太守乐庭瑰
供养群像服饰艺术
再现
敦煌莫高窟盛唐第130窟

晋昌郡太守乐庭瑰供养群像包括多位男供养人像和侍从像，这里选取其中的四身人物像进行服饰艺术再现。这组供养人像均穿着唐代典型的官服，即软脚幞头、圆领袍、革带和乌靴的组合，服饰色彩以石绿色、绛红色、茜色和白色为主，主次分明，错落有致，表现了众人尊贵有序的身份和一心供养的虔诚。

The painting of Prefect Yue Tinggui of Jinchang county worshiping group include many male donors and attendants. Here, four of them are selected for clothes art reproduction. This group of donors are all dressed in the typical official clothes of the Tang Dynasty, which are, soft ends Fu hat, round collar robe, leather belt and black boots. The dress colors are mainly malachite green, crimson, alizarin and white, with clear priorities and well balanced, showing their identity in order and the piety by wholeheartedly offering.

The artistic reappearance
of Prefect Yue Tinggui's
clothes of Jinchang county
in Cave 130 of the high Tang Dynasty at
Dunhuang Mogao Grottoes

晋昌郡太守乐庭瑰供
养像服饰艺术再现
敦煌莫高窟盛唐第130窟

晋昌郡太守乐庭瑰手持长柄香炉，容貌俊朗，浓眉凤眼，胡须飘然。他头戴黑色软脚幞头，身穿绿色暗纹提花圆领襕袍，极具丝质材料的垂感和飘逸感。腰系黑色革带，侧面斜插笏板，脚上着黑色软靴。其装束具有唐代典型的官府公职服饰特征。

Prefect Yue Tinggui of Jinchang county holding a long handled incense burner, looks handsome, with thick eyebrows, phoenix eyes and long beard. He wears a black soft ends Fu hat and a green hidden jacquard round collar Lan robe, which has a great sense of draping and lightness of silk materials. The waist is tied by black leather belt, a Hu board is obliquely inserted on the side, and the feet in black soft boots. His clothes have the typical official clothes characteristics of the Tang Dynasty.

The artistic reappearance of
the Prefect's wife worshiping
group clothes
in Cave 130 of the high Tang Dynasty at
Dunhuang Mogao Grottoes

都督夫人供养群像
服饰艺术再现
敦煌莫高窟盛唐第130窟

与晋昌郡太守乐庭瑰供养群像相对的是其夫人、女儿及侍女的供养像，这里选取最为主要的五身人物像进行服饰艺术再现。整体服饰造型雍容华贵，色彩绚烂夺目，其艺术表现风格与传世的唐人绘画名作《簪花仕女图》和《捣练图》有异曲同工之妙，体现出盛唐时期以丰腴为美的审美取向。

On the opposite of the worshiping group images of Prefect Yue Tinggui of Jinchang county are the worshiping images of his wife, daughters and maids. Here we choose the most important five figures to make the clothes art reproduction. The overall dresses are elegant in shape and gorgeous in color. Their artistic expression style is similar to that of the handed down famous paintings *Group Portrait of Noble Women* and *the Picture of Making Clothes* of the Tang Dynasty, which reflect the aesthetic preference of plump as beauty in the high Tang Dynasty.

The artistic reappearance of
Prefect's wife clothes
in Cave 130 of the high Tang Dynasty at
Dunhuang Mogao Grottoes

都督夫人供养像服饰
艺术再现
敦煌莫高窟盛唐第130窟

走在最前列的是都督夫人太原王氏，她身着盛装，面容圆润优美，画桂叶眉，凤眼丰唇，束高耸的峨髻，髻上插饰花钗和梳篦。都督夫人身穿绿色交领宽袖短襦，外罩绛红色底花半臂，下穿红色曳地长裙，肩披米白色披巾，腰系绿色襜襦，脚穿笏头履。其服饰图案以团花纹、折枝花纹、花叶纹为主，以印花和刺绣工艺加以呈现，结合大胆的配色，显得格外华美艳丽。

At the forefront is the Prefect's wife who came from Taiyuan Wang family. She dresses gorgeously, with a round and beautiful face, laurel leaf eyebrows, phoenix eyes and full lips, and tied in high E'ji hair bun which inserted with flower hairpins and combs. She wears a green cross collar, wide-sleeved short Ru, a crimson half-sleeved coat decorated with flower pattern, a long red skirt, a yellow shawl on shoulders, green brocade Xianli on her waist, and Hu Head shoes on feet. Her dress patterns are mainly round flower pattern, small branches pattern and flower leaf pattern, which are presented by printing and embroidery techniques, and combined with bold color matching, looks marvellous.

The artistic reappearance
of offering image of the
eleventh daughter's clothes
in Cave 130 of the high Tang Dynasty at
Dunhuang Mogao Grottoes

女十一娘供养像服
饰艺术再现

敦煌莫高窟盛唐第130窟

位于都督夫人身后的是她的两个女儿，其一榜题为"女十一娘"。她和母亲的妆容发型相似，也画桂叶眉，脸上点饰有面靥，束高耸的峨髻，髻上也装饰着花钗和梳篦。女十一娘双手持花束，上穿红色交领宽袖短襦，下着绿色落地长裙，腰系红色襳褵，肩披白色披巾，脚穿五朵履，整体服饰色彩对比鲜明、华美浓丽。其服饰图案风格与其他几身女供养人像相似，均以散点状的花卉植物纹为主，体现了当时世俗化的审美观念。

Behind the prefect's wife are her two daughters, one has inscription "the eleventh daughter". Her makeup and hairstyle are similar to that of her mother，also has laurel leaf eyebrows, and the face is decorated with Mianye. High E'ji hair bun, which is also decorated with flower hairpins and combs. The eleventh daughter holds a bouquet by both hands, wears a red cross collar wide-sleeved short Ru on the upper body, and a green long trained skirt, tied a red Xianli around waist, a white silk shawl on her shoulders and five petal-shaped-heads shoes on her feet. The overall dress is bright and gorgeous, and the dress pattern style is similar to other female figures, which are mainly dots clusters flower and plant patterns, reflecting the secularized aesthetic concept at that time.

The artistic reappearance
of offering image of the
thirteenth daughter's clothes
in Cave 130 of the high Tang Dynasty at
Dunhuang Mogao Grottoes

女十三娘供养像服
饰艺术再现
敦煌莫高窟盛唐第130窟

这是都督夫人的另一位女儿，壁画题榜标注为"女十三娘"。她的面容丰满圆润，画桂叶眉，丹凤眼，脸部点饰面靥，头戴凤冠，斜插步摇，同时饰花钿和角梳。女十三娘身穿米白花色短襦，外披绿色半臂，下着织花长裙，肩披淡绿色披帛，腰间垂红色织锦襳褵，脚穿翘头履。她的服饰图案以散点状的叶片纹、折枝花纹为主，与都督夫人供养像相比，服饰色彩较为淡雅，透露出盛唐时期偏爱植物和崇尚自然的审美趣味。

This is another daughter of the prefect's wife, and the inscription is "the thirteenth daughter". Her face is plump and round, with laurel leaf eyebrows, phoenix eyes, Mianye, and a phoenix crown on her head, which inserted with Buyao(a kind of hairpins), and small flowers and combs. The thirteenth daughter wears a short Ru in rice white inside, a green half-sleeved coat outside, the lower body wears a long jacquard skirt, light green silk shawl on her shoulders, red brocade Xianli around her waist, and upturned heads shoes on her feet. Her dress patterns are mainly doted leaf pattern and small branches pattern. Compared with the worshiping image of the Prefect's wife, the dress color is simpler, revealing the aesthetic taste of preferring plants and advocating nature in the high Tang Dynasty.

The artistic reappearance
of offering images of
maids' clothes
in Cave 130 of the high Tang Dynasty at
Dunhuang Mogao Grottoes

侍女供养像服饰艺
术再现

敦煌莫高窟盛唐第130窟

　　都督夫人供养群像中共有九身侍女像，这里选取其中的两身进行服饰艺术再现。两位侍女梳双垂髻，面容稚嫩，手托花盘供养。她们均着男式的圆领袍服，束革带，穿乌靴，体现了盛唐时期女着男装的时尚潮流。其中一身侍女着石绿色袍服，上有刺绣的菱形散花纹，另一身侍女着红色圆领袍，采用朵花暗纹织物面料制作，两身人物服饰色调一冷一暖，图案一花一素，体现了一静一动、相得益彰的人物性格。

The prefect's wife worshiping group images have nine maids. Here, two of them are selected for clothing art reproduction. The two maids combed their hair into Shuangchui hair bun, and their faces are young and holding flower trays by both hands. They all wear men's round collar robes, leather belts and black boots, reflecting the fashion trend of women wearing men's clothes in the high Tang Dynasty. One maid is dressed in a malachite green robe embroidered with diamond shaped flower clusters pattern, and the other maid is dressed in a red round collar robe made of hidden flower jacquard material. The colors of the two figures' clothes are one cold and one warm, and the patterns are one complex and one simple, reflecting the personality of one static and one dynamic characters which complement each other.

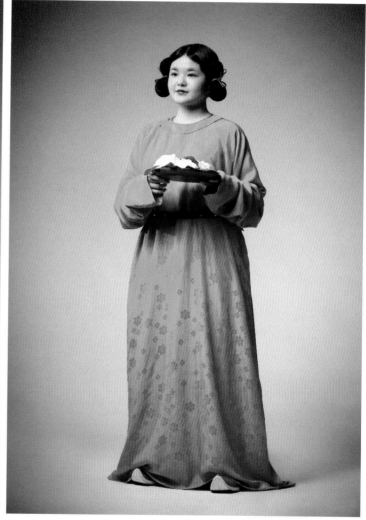